Passive Solar Energy

The Homeowner's Guide to Natural Heating and Cooling

Bruce Anderson and
Malcolm Wells

Illustrated by Malcolm Wells

 BRICK HOUSE PUBLISHING CO.
Andover, Massachusetts

This book is dedicated to
Deborah Napior
my best friend and compatriot
B.A.

Cover Photo: Douglas Balcomb
Interior view of the large solar room of Douglas and Sarah Balcomb's house in Sante Fe, New Mexico. Not only are their annual heating and cooling bills less than one hundred dollars, but the quality of their lives as a result of their energy-conserving, passive solar home is among the highest. (Designed by Sun Mountain Design, construction and engineering by Communico, especially Wayne and Susan Nichols.)

Illustrated by Malcolm Wells

Published by Brick House Publishing Co., Inc.
34 Essex Street
Andover, Massachusetts 01810

Production credits:
  Editor: Jack Howell
  Book design: Dixie Clark/Ned Williams
  Cover design: Ned Williams
  Typesetting: Neil W. Kelley
  Production supervision: Dixie Clark

Printed in the United States of America

LIBRARY OF CONGRESS

CATALOG CARD NO.: 80-70147
    Anderson, Bruce etals
        Passive solar energy

    Andover, Mass : Brick House Publishing Co.

    194 p.
    8012

801023

Contents

Foreword v   Preface viii
Acknowledgements ix

Part 1 Solar Basics   1

Chapter 1: Why Passive?   3

I Thought You Said It was Simple 6
What, More Definitions? 10

Chapter 2: Solar Building Design Basics   18

Solar Retrofit 24
The Ins and Outs of Energy Flow 26
Solar Position 30
The Site 33

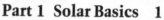

Part 2 Passive Solar Heating   39

Chapter 3: Solar Windows   41

Thermal Mass 44
Movable Insulation 49

Chapter 4: Solar Chimneys   52

Basic System Design 53
Systems with Heat Storage 55
Air Flow 56
Costs and Performance 59

Chapter 5: Solar Walls   60

Costs 62
Construction 66
Converting Your Existing Home 69

Chapter 6: Solar Roofs   70

Chapter 7: Solar Rooms   73

Heat Storage 75
Costs 77
Large Solar Rooms 77

Part 3 Passive Solar Cooling   85

Chapter 8: How the Sun Can Cool   87

Solar Control 87
Ventilation 90
Evaporation 94
Radiational Cooling 95
Ground Cooling 96

Part 4 Putting the Pieces Together 99

Chapter 9: House Design Based on Climate 101
  A Tradition of Regional Architecture 101
  A Revival of Climate-Based Architecture 102

Chapter 10: The Hard Choices of Cost 110
  How Large 110
  What Does Passive Cost? 114

Afterword 118

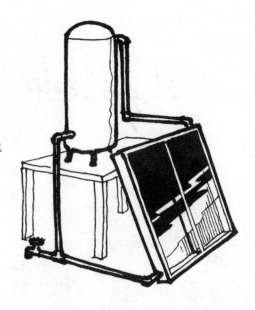

Photo Section 119
  Solar Windows 119
  Solar Chimneys 121
  Solar Walls 122
  Solar Rooms 124
  Solar Roofs 126

Construction Details 127

Appendix 1: Sun Path Diagrams 141

Appendix 2: Solar Radiation on South Walls on Sunny Days 144

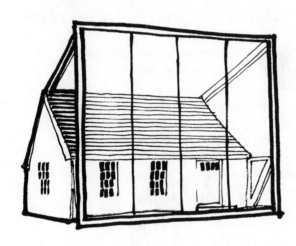

Appendix 3: Maps of the Average Percentage of the Time the Sun Is Shining 148

Appendix 4: U-Values for Windows with Insulating Covers 153

Appendix 5: Degree Days and Design Temperatures 154

Appendix 6: Selecting the Right Glazing Material 160

Directory of Passive Resources 165

Bibliography 192

Index 195

Foreword

A sense of apprehension permeates our energy debate. Will there be another oil shortage? Will OPEC and the oil companies tighten their grip on our pocket books? How long can our economy survive an endless series of oil price increases?

As I have travelled this nation this last year, I have discovered that despite the sense of hopeless and helpless frustration that too often dominates the energy debate in Washington, there is a revolution of Yankee ingenuity changing the energy future of local communities.

I have seen the vision of that future—that rebirth of Yankee ingenuity—from Maine to California. In New England, millions of barrels of oil have been saved because hundreds of thousands of citizens have turned to wood stoves to heat their homes. Builders are selling homes so energy efficient that they need no furnaces.

In the farm states, gasohol production has expanded 20 times in the last two years alone.

In the Tennessee Valley, tens of thousands of solar water heaters have been installed as part of an innovative TVA program.

In the Southwest, most new homes are using some solar energy; and in California, whole communities have been designed to tap the solar potential.

While officially ignoring this quiet revolution of energy independence, government has been spending billions of additional dollars on breeder reactors and oil shale extraction. So far, this spending has done almost nothing to cut America's overreliance on OPEC; but solar energy and energy efficiency have already achieved major reductions in oil imports. Improvements in energy efficiency since 1973 have saved the equivalent of three and a half million barrels of oil a day—two and a half times as much as the contribution from increased domestic

v

production. All forms of renewable energy account for the
equivalent of 2.7 million barrels a day, or twice the amount
of oil flowing through the Alaska Oil pipeline.

The quiet revolution has its quiet revolutionaries—
Americans with the personal initiative of a Bruce Anderson
and a Malcolm Wells.

In *Passive Solar Energy*, Anderson and Wells, two of the
foremost experts in the solar field today, have written a
comprehensive reference on passive solar heating and
cooling, the most accessible and least expensive form of
renewable energy now available to consumers. As the
authors point out, passive solar has all too often been
overlooked in the rush to develop and install active solar
collectors. Only recently have we learned that collectors are
not necessarily the right solar system for every home; in fact,
about one-half of the solar homes in the United States collect
solar energy without special collector systems.

Passive Solar Energy replaces much of the technical jargon
which pervades energy discussion with easy-to-
understand terms. Instead of "thermal storage walls," we
read of a "solar wall" that can collect heat during the day and
release it slowly and evenly through the night. For
"convective loop," a dense term that conjures up images of
twisted piping, the term "solar chimney" is used to describe
a solar collector attached to the south wall of a home.

As such simplified terms imply, part of passive solar's
appeal lies in its elegant simplicity. It can be nearly as easy to
add passive solar features to a home or business as it is to
describe them.

I was on hand to dedicate the most recent MIT solar house
on Sun Day in May, 1978. Its most notable feature is the use
of what Anderson and Wells call "heat-of-fusion storage
material"—meltable salts which store and release heat with
only small changes in temperature. With wider
understanding of such innovative technology, and greater
federal support, I believe passive solar will become one of the
first renewable forms of energy to make a substantial
contribution to heating our homes while ending our
overdependence on the OPEC cartel.

It is time for this nation to hear and heed the words of the
master teacher of Western civilization, who wrote twenty-
four centuries ago:

Now in houses with a south aspect, the sun's rays penetrate into the porticos in winter, but in the summer, the path of the sun is right over our heads and above the roof, so that there is shade. If then this is the best arrangement, we should build the south side loftier to get the winter sun and the north side lower to keep out the winter winds. To put it shortly, the house in which the owner can find a pleasant retreat at all seasons and can store his belongings safely is presumably at once the pleasantest and the most beautiful.

Socrates
Xenophon's Memorabilia

Senator Edward M. Kennedy
September, 1980

Preface

Some skeptics say solar energy won't work. But you know better; you've been using it all your life. Think of how nice and toasty your sunporch gets. And the way you can almost bake bread inside your car on a sunny October day. Remember how your favorite begonias had sunstroke in the greenhouse last April? There's a lot of heat in all that light.

It's only natural, then, that thousands of people are turning to solar energy to heat their homes. Many of the countless newspaper reports focus on fancy solar homes of the future, replete with lots of "active" moving parts: collectors, pumps, fans, valves, heat exchangers, and electronic controls. "Active" systems can save a lot of energy (and are certainly preferable to systems that use coal, gas, oil, and nuclear power!), but it's no secret that they are often needlessly complicated and expensive.

Solar heating and cooling doesn't have to be complicated and expensive. This book is for all of us who know that solar energy works—who have burned our behinds on the sun-baked seats of our cars, who have been burnt by uncertain energy supplies and skyrocketing prices, and who are ready to do something about it all—just as soon and as simply as possible.

"Passive" solar heating and cooling does not depend on pumps or fans or any other devices. Instead, it relies on the natural ebb and flow of the energy of the sun through a house. With a few facts from this book and a little common sense, you can combine passive solar design with energy conservation and reduce the heating and cooling bills for a new house to less than 15% of those for conventional houses. And many of the ideas can be adapted to existing houses as well.

It's no wonder, then, that more and more people who are planning to build houses someday are thinking *passive solar*. It is the natural first step toward living better while using less energy. Those who live in passive homes bathed in winter sunshine wish that everyone else could learn how to capture more warm solar energy.

We won't discuss here the pros and cons of *active* systems. Dozens of excellent books already do that. We have enough to do in presenting the passive story on these pages. Our aim is to make this presentation as simple as the systems it describes. So, here it is. We hope it will be helpful.

Bruce Anderson & Malcolm Wells

Acknowledgements

Few books are the result of the efforts of only one or two people. *Passive Solar Energy* is no exception. On the contrary, although but two authors are given prominent visibility, this book is the product of a unique and memorable set of people and circumstances.

The term "passive" was not applied to "solar energy" until six years ago or so when it was "invented" by Benjamin "Buck" Rogers, a venerable engineer from northern New Mexico. An associate of Buck's at Los Alamos Scientific Laboratories, Dr. J. Douglas Balcomb, led many of the scientific and technical analyses which confirmed the performance and clarified appropriate applications of passive systems.

Much of this and other research, development, and market work has been funded by the Department of Energy. Michael Maybaum, the manager of DOE's passive programs, has been a particularly strong supporter of passive.

But the real credit for getting passive moving in this country goes to people—not government. Several dozen dedicated men and women knew that passive made sense long before it became fashionable, and they risked their professions and fortunes to prove it. In a few years, this dedicated core grew to several thousand: inventors, scientists, engineers, architects, manufacturers, builders . . . and buyers.

In preparing this book, my greatest thanks goes to my co-author Malcolm Wells. Mac's astute wit and wisdom, pithy contributions, and scintillating drawings breathe life into an otherwise uneventful text. The foreword is representative of Senator Edward M. Kennedy's faith in the innate potential of the individual. His relentless efforts to help the common man and woman tap that potential are paralleled by few other people.

A number of people and organizations provided photographs, drawings, or information. They are too numerous to mention here but are given credit at appropriate locations in the book. However, I would like to mention two organizations whose staffs contributed significantly. Total Environmental Action, Inc. provided construction and technical information throughout. *Solar Age* magazine, published by SolarVision, Inc., provided much information, especially for the resource directory. Both companies are based in Harrisville, New Hampshire.

Not all authors are as fortunate as I in having support and cooperation. In particular, the staff of Total Environmental Action Foundation, Inc., Harrisville, N.H., provided timely and pivotal assistance. Bill Frye, TEAF's Director, contrib- uted extensively to the conservation and retrofit sections and to the resource directory. Jennifer Harris edited the book, coordinated preparation of the construction details, pulled to- gether the photographs, and helped with dozens of other details.

Rick Katzenberg and Deborah Napior also reviewed and edited the manuscript, helping enormously to clarify muddy passages. Peter Harrison transformed visually esoteric drawings into 3-D clarity, which helped Mac to prepare the construction details. Julie Carothers, Karen Weiner, and Barbara Hann typed various drafts. Debbie Doscher and Mary Mason proofread galleys. The production team led by Dixie Clark nearly created tornadoes as they met impossible deadlines. Ned Williams designed the beautiful cover and the book. Jack Howell, publisher, and Jim Bright of Brick House organized and coordinated this entire "event." Rick Katzenberg, my good friend who supported me in so many ways in the publishing of my first book, *The Solar Home Book*, did so again in countless more.

Finally, I dedicate this book with gratitude to Deborah Napior, my best friend and compatriot, whose enduring patience supported me during tedious hours of writing.

May you the reader respond to this book with new awareness of the vast potential of energy conservation and passive solar energy, and with action to make tomorrow in the image of this awareness.

Bruce Anderson
Harrisville, New Hampshire
October 1980

PART 1
Solar Basics

Ancient Native American cliff dwellings in the southwest were designed and built in response to the sun. (Photo by Malcolm Lillywhite, Lakewood, Colorado.)

Chapter 1

Why Passive?

Passive solar designs are simple. This simplicity means greater reliability, lower costs, and longer system lifetimes.

Since passive systems have few (if any) moving parts, they perform effortlessly and quietly without mechanical or electrical assistance. Simplicity lowers the cost of the job. Without motorized dampers, automatic valves, sophisticated control systems, or high-tech components, much of the work can be done using standard building materials and basic construction skills.

The most significant reason that passive makes sense economically is that most passive designs are inherently durable, lasting at least as long as the rest of the house with little or no maintenance or repair. Conventional building materials such as glass, concrete, and brick weather well and are generally longlasting. For the life of the house, a passive system should continually maintain, if not improve, its value at least as well as the rest of the house. It should require little more maintenance than a standard wall or roof.

Because you can build passive designs in small sizes, the initial effort need not involve a large financial commitment. Instead, the first step can be relatively small with correspondingly little risk.

For optimum performance, some passive systems require daily or monthly adjustments of shades, shutters, or vents. Although some people may at first regard this as an imposition, it is really no more trouble than operating a dishwasher or closing draperies in the evening. Before long, passive-home residents will find these to be pleasant routines that bring them closer to the flux of the

environment to which their homes are atuned. They are usually rewarded with a rich and exciting living experience as a result of their efforts, while saving both energy and money.

Radiant heat from large passive collecting surfaces is usually more comfortable than the drafty heat of conventional hot air or hot water heating. In well-designed systems, temperature variations are small, generally within a range of 5° to 10° each day. But in less well-designed houses, temperatures can vary more widely. Some solar enthusiasts feel such temperature fluctuations are natural, and not uncomfortable, particularly at the higher end. In fact, many passive home residents enjoy the warmer-than-usual temperatures on a sunny winter day.

Passive solar systems save fossil fuels. The economy is benefitted because the nation imports less oil. And since passive energy systems do not require transmission lines, pipe lines, or strip mines, they produce neither dangerous radioactive wastes nor polluted air and water. Passive systems have few negative consequences. They can use renewable and recyclable materials, and they produce jobs.

If a house has low heating or cooling requirements, and if a passive solar system is designed to provide only a small fraction of the energy, the system can be small and have only a slight effect on the overall appearance of a house. It need not make the house unattractive. In fact, properly designed passive houses can be more beautiful than conventional ones. Picture it—large expanses of south-facing glass overlooking your yard; a beautiful sunspace filled with plants year 'round. You can save energy, save money, and provide a better living environment, all at the same time! Comparing a good passive house to a conventional one is like comparing a modern, dependable lightweight bike to the high-wheeled terrors of the 1890s.

I Thought You Said It Was Simple

It is. Every material and principle incorporated into passive solar design is common and in everyday use. The melting of an ice cube or the ability of a stone to stay warm long after sunset—these are the kind of considerations on which all passive design is based. The only trick is to learn the labels so that it is easier to understand and discuss. Then you can say "thermal mass" instead of having to say (each time you discuss the phenomenon) "the ability of a stone to stay warm long after sunset."

Conduction
The transfer of heat between objects by direct contact.

Natural Convection
The movement of heat through the movement of air or water.

Thermal Radiation
The transfer of heat between objects by electromagnetic radiation.

Mean Radiant Temperature
The average temperature you experience from the combination of all of the various surface temperatures in a room—walls, floors, ceilings, furniture and people.

Air Stratification

The tendency of heated air to rise and to arrange itself in layers with the warmest air at the top.

Degree day

A unit used to measure the intensity of winter. The more degree days there are in total for the season, the cooler the climate.

Windows

Windows let light (and heat) in (and out).

Evaporative Cooling

Natural cooling caused by water's ability to absorb heat as it vaporizes.

Insulation

Materials that conduct heat poorly and thereby reduce heat loss from an object or space.

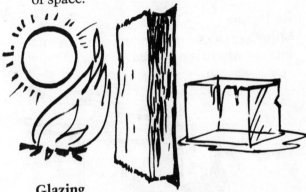

Glazing

Layers of glass or plastic, used in windows and other solar devices for admitting light and trapping heat.

Shading

Measures for blocking out unwanted sunlight that can overheat the house.

Movable Insulation

Insulating curtains, shutters, and shades that cover windows and other glazing at night to reduce heat loss.

Reflectors

Shiny surfaces for bouncing sunlight or heat to where it's needed.

Thermal Mass

Materials that store heat. Heavy materials (concrete, stone, and even water) store a lot of heat in a small volume, compared with most lightweight materials, and release it when needed.

Heat-of-Fusion Storage Materials

Meltable materials store heat when they "phase change" from solid to liquid form and release heat when they re–solidify. They require less mass (and volume) to store the same amount of thermal energy as more conventional heat storage materials, and only small changes in temperature are necessary to induce the phase change.

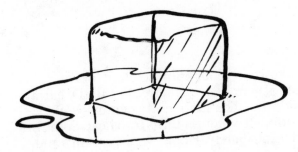

R-Value

A measure of the insulating ability of a material, a wall, a ceiling, etc. The higher the R-value, the better the insulation and the less the heat loss.

U-Value

A measure of the rate of heat loss through a wall or other part of a building. It is the reciprocal of the total R-values present. The lower the U-value, the lower the heat loss.

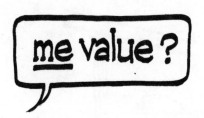

What, More Definitions?

No, just a preview. The following are the passive systems
covered by this book. As with everything else, there are
both advantages and disadvantages. For most of us,
though, the advantages far outweigh the disadvantages.

1. Solar Windows

When you make a conscious effort to place lots of glass
on the south side of your house, feel free to call the extra
glass "solar windows." The sunlight that enters your
house directly through windows turns into heat. Some of
the heat is used immediately. Floors, walls, ceilings and

furniture store the excess heat. Movable insulation can cover the windows at night to reduce heat loss. South glass takes advantage of the winter sun's low position in the sky. In the summer, when the sun is high, the glass is easily shaded by roof overhangs or trees. Solar windows are often referred to as "direct gain" systems.

Advantages

Everyone can use this simplest of all solar designs. In fact, most of us already do, but not nearly as much as we should. Solar windows are inexpensive—often free—and they provide a light and airy feeling.

Disadvantages

Not everyone appreciates sunshine pouring in all day. Many people enjoy the extra heat and the higher temperatures, but sunlight can fade fabrics, and too much glass may cause too much glare.

2. Solar Chimneys

Air is warmed as it touches a solar-heated surface. The warmed air rises, and cooler air is drawn in to replace it. This is what happens in an ordinary chimney. The process of natural convection can occur in a continuous loop between your house and a solar collector attached to its south wall. As the air in the solar collector is heated, it expands, rises, and enters the house. Cooler house air is drawn into the collector to take its place. This is why these "solar chimneys" are usually referred to as "convective loops." Before too long, you should be able to buy solar chimney collectors from your local solar retail outlet or solar installer.

Advantages

Solar chimneys are very simple and avoid many of the problems of direct gain systems, such as glare and heat loss. Also, they're easy to attach to present homes.

Disadvantages

Like direct gain, too large a system may result in higher than normal temperatures in your house. Careful construction is required to ensure proper efficiency and durability.

3. Solar Walls

When the mass for absorbing the sun's heat is located right inside the glass, you have a "solar wall." The wall, painted a dark color, heats up as the sun passes through the glass and strikes it. Heat is then conducted through the wall and into the house.

Another type of solar wall substitutes water for masonry. Tall cylinders of water, 55-gallon barrels, and specially-fabricated water walls are common. The water containers radiate their solar heat directly to the room. These walls are often referred to as "thermal storage walls."

Advantages
Thermal storage walls have many of the same advantages as convective loops and simultaneously solve the heat storage problem. The mass is right where it belongs—in the sun. The thermal mass keeps the house at pretty even temperatures nearly 24 hours per day.

Disadvantages
The wall also loses heat back to the out-of-doors through the glass. Triple glazing or movable insulation solves this problem in cold climates but can be costly. Keep in mind that construction of the wall can be expensive and may in some cases reduce available floor area.

4. Solar Roofs

"Solar roofs" are like solar walls, only guess where the heat storage is instead! They are often called "thermal storage roofs." Most solar roofs use water in large black plastic bags (like waterbeds) to absorb heat during the day. The water ponds store the heat, which is in turn conducted through the ceiling and radiated to the house below. Insulating panels cover the ponds at night to reduce heat loss. *

Solar roofs can, in certain climates, cool your house during the summer; the water absorbs heat from the house below and radiates it to the cool night sky. Insulating panels shade the ponds by day.

---

* The most widely known solar roof, developed by Harold Hay, is called "Skytherm."

Advantages

These systems can, in some climates, provide for all your heating and cooling needs. And they can do so while keeping you as comfortable, or even more comfortable, than almost any other type of heating system, whether solar or conventional.

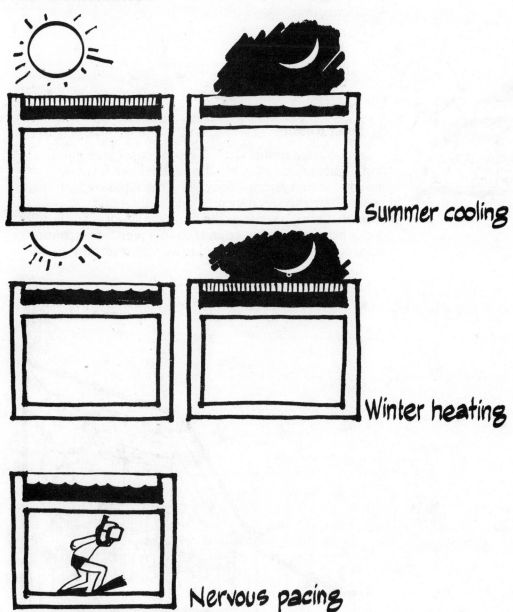

Summer cooling

Winter heating

Nervous pacing

Disadvantages
Solar ponds require careful design, engineering and construction. Although little information is presently available, further research and development is currently under way. Their efficiency and cost effectiveness are not nearly as good in cold climates as they are in dry, sunny, southern ones.

5. Solar Rooms

Solar-heated rooms such as greenhouses, sun porches, and solariums are possibly the hands-down favorite passive solar system. They give the house extra solar-heated living space; they provide a feeling of spaciousness—a sense of the outdoors; they act as buffer zones between the house and outdoor weather extremes. Solar rooms are often referred to as "attached sunspaces."

Advantages

Solar rooms can greatly improve the interior "climate" of a house. Although solar room temperature swings may be large, since plants can tolerate much wider swings than people can, house temperature swings can still remain small (3°–8°). Solar rooms can add humidity to the house air if desired. A solar room can become additional living space for a relatively low cost. Besides, people love them! And they are readily adaptable to present homes.

Disadvantages

Improperly designed or built solar rooms will not work well. Although construction costs can be kept down, good quality construction is expensive. Factory-built kits can also be expensive.

Chapter 2
Solar Building Design Basics

To most people, energy conservation lacks the glamour of solar energy, but it is always a winner in saving energy. In any climate where heating or cooling is a "big thing," solar design done in combination with energy conservation works best. Conservation always pays off in savings faster than any other energy strategy. The whole idea is simple.

A tight, energy-conserving, passive solar home may reduce energy costs by 50 to 90 percent depending on climate. As costs of conventional energy soar and fears of interrupted supplies heighten, your wise investment becomes more obvious to everyone: yourself, your friends, and . . . the next buyer of your home.

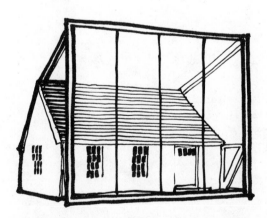

South glazing sized to heat a poorly-insulated house.

Much of the information in this section is based on *Solar Home Heating in New Hampshire* (available through the Governor's Council on Energy, 2½ Beacon Street/Concord NH 03301)

Federal tax credits of 15 percent make investments in conservation very attractive. I.R.S. publication No. 903 "Energy Credits for Individuals," will tell you how the government endorses your investment by giving a 15 percent "discount" that you can subtract directly from your tax bill.

Basking in the winter sun can be pleasant even though the temperature is below feeezing. But you would never go out without "buttoning up" first. It should be just as obvious that it makes no sense to leave your house out in the cold without buttoning it up first. Most heat is lost through either conduction or infiltration. Conduction losses occur as heat escapes through the roof and walls. Infiltration means losing heat through warm air escaping and being replaced by cold air drafts seeping into the house through cracks around doors, windows, and around the foundation.

In climates where energy is used for cooling, conservation comes first, too! People do not sit in the hot sun for hours without shielding themselves. So also, shading, ventilating, and insulating to keep heat out and cool air in are important to summer cooling. These simple, economic measures lower both the size and cost of air conditioning equipment, reduce cooling bills, and improve comfort by lessening the often large differences felt between indoor and outdoor temperatures and humidity levels when air conditioning is used.

In terms of costs, energy conservation measures fall into three categories: Free, Cheap, and Economic. You just can't miss. Let's talk about some specific energy conservation measures, starting with the free ones.

Free heating energy conservation measures include lowering thermostat settings to 68° or lower during the day and 55° or lower at night, reducing hot water tank temperatures to 120° (or 140° if dishwasher instructions require), adding water-saving shower heads, closing chimney dampers and blocking off unused fireplaces, shutting off unused rooms, turning off lights, and wearing heavier clothing. For cooling, get used to slightly warmer temperatures or turn the air conditioner off altogether and open windows. When these measures are used in combination, 20 percent savings is easily obtainable at no cost.

Among the "cheap" energy conservation measures are maintaining the efficiency of your heating system through servicing check-ups, caulking and weatherstripping windows and doors to seal infiltration cracks, installing a clock thermostat for automatic set-back, and adding sheets of plastic to windows that are the big heat losers. For cooling, shades, awnings,—and trellises with plant growth—block out the sun. Fans are much less expensive to run than air conditioners.

"Economic" energy conservation measures include adding extra insulation in attics and walls and around foundations, adding storm windows and doors or replacing older windows with tight new ones, covering windows with night insulation, and adding a vapor barrier wherever possible. Replacing an old inefficient burner or furnace can be economic. Consider a woodburning furnace or stove. Add an airtight entryway or plant a wind break of trees. (But don't block out south sunlight!) These measures do require an initial dollar investment, but they almost always make economic sense, even if a bank loan is necessary to finance the investment.

So far, we have talked about the economics of energy conservation primarily for retrofitting existing homes. Planning energy conservation into new house design is a whole "new game" economically, one in which you stand to reduce energy bills even further. (Remember, energy conservation is the first step to efficient solar heating!) By planning before construction, many of the "economic" measures become "cheap" or "free." Extra insulation and a vapor barrier, for example, are very inexpensive in new construction as compared to retrofitting with the same materials. The small extra cost of "doing it right" may be offset by lower construction costs of other items, such as a smaller furnace, as well as by energy savings.

You are almost ready to do "something solar" to your home. But first, we want to elaborate upon some options. They are presented in their relative order of cost effectiveness.

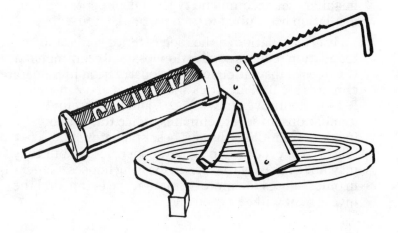

Caulking and Weatherstripping

Caulking and weatherstripping reduce air infiltration by sealing cracks around windows, doors, wall outlets, and the foundation. These materials are inexpensive and easy to use. Caulkings include silicones, urethanes, and materials with an oil or latex base. Weatherstripping is made of felt, foambacked wood, and vinyl and steel strips. A supplier can inform you of types, costs, life expectancies, and uses of these materials. One- to-five year paybacks are usual, but sealing off big cracks may pay for itself in a matter of months.

Insulation

Insulating materials are assigned "R-values," a rating of how well each material resists the conduction of heat energy. The higher the R-value per inch thickness of material, the more effective the insulation.

Insulation types include fiberglass (which is available in a range of thicknesses), bags of loose cellulose, blow-in foams, and plastic foam sheets or boards. Check with reputable suppliers for possible fire or health hazards of these materials as well as for their comparative durability. Installers can recommend types and amounts of insulation best suited to your particular house.

Heavy insulation in the roof, exterior walls, and foundation reduces conduction losses. Recommended R-values of insulation for new construction in moderate climates are R-38 in attics and ceilings, R-19 in walls, R-19 in floors over crawl spaces, and R-11 around foundations. In severe climates, twice this much insulation should be used. These standards should be followed or exceeded in all new construction, but they may not be as easy to reach in existing houses. Generally, in older houses the more insulation the better, and the investment will be a good one.

Vapor Barriers

Vapor barriers protect wall and ceiling insulation from moisture. As warm house air seeps through walls and ceilings to the cold outside, it carries moisture with it into the insulation. If moisture condenses and is trapped, the insulation loses much of its effectiveness. In severe cases, moisture can cause wood to rot.

Adding vapor barriers to existing houses is difficult unless interior walls have been torn down. However, interior vapor barrier paints and vinyl wall papers are available for existing houses. Ridge and soffit vents for attic ventilation help carry away moisture. In humid climates, consult local builders since vapor barriers are not used in some parts of the country where moisture moves from the outside into walls during the summer.

Storm Windows and Doors

A window with a single pane of glass (called single glazed) can lose 10 times the heat of an R-11 wall (3½" of fiberglass insulation) and 20 times the heat of an R-19 wall (6" fiberglass insulation). Adding a second layer of glazing can save one gallon of oil each winter per square foot of window area in cold climates. Adding two layers (resulting in triple glazing) can save nearly two gallons of oil per square foot of window area. At 1980 energy prices, added glazing can save $10 to $15 per window each year. Storm

Human Comfort

Regardless of how fuel bills are reduced, the primary purpose of energy consumption for heating and cooling is to keep people comfortable. Passive solar design is a natural strategy for accomplishing this.

Our bodies use three basic mechanisms for maintaining comfort: convection; evaporation/respiration; and radiation. Air temperature, humidity, air speed, and mean radiant temperature all influence how we use our comfort control mechanisms.

Perhaps mean radiant temperature is least understood. Mean radiant temperature (mrt) refers to the average temperature we feel as a result of radiant energy emanating from all surfaces of a room: interior walls, windows, ceilings, floors, and furniture. It combines with the room air temperature to produce an overall comfort sensation, and different combinations of mean radiant temperature and room temperature can produce the same comfort sensation. For example, if the air temperature is 49° and the mean radiant temperature is 85°, you will feel as though it is actually about 70°. The same holds true for the combination of an air temperature of 84° plus a mean radiant temperature of 60°.

Many passive systems use warm surfaces to keep a house comfortable. The higher mean radiant temperatures provide comfort at lower air temperatures. Most people prefer this comfort balance to the more common comfort balance found in conventional houses where warm air is surrounded by cool or cold surfaces. In other words, because you are surrounded by warm surfaces, a passive house makes you "feel" warmer even with room temperatures several degrees lower than you might have in a conventional house.

Once you've become accustomed to passive warmth, conventionally-heated rooms feel cool and drafty, even at identical air temperatures. Interior surfaces of thickly insulated walls, floors, and roofs are warmer than those that are poorly insulated. The same holds true for multipaned windows compared with single glazed. Thus, energy conservation enhances mean radiant temperature and is a good companion to passive solar in providing comfort.

Because lower house temperatures result in less heat loss, even more energy can be saved than what is normally calculated.

The combinations of temperatures in the following chart produce the common comfort sensation of 70°F:

Mean Radiant Temperature:	85	80	75	70	65	60	55
Air Temperature:	49	56	64	70	77	84	91
Comfort Sensation:	70	70	70	70	70	70	70

windows are available with wood, aluminum or plastic sash. Aluminum insulates least and wood insulates best. Storm windows come with single or double panes of glass. Placing clear plastic sheets over windows is the least expensive solution.

Thermal Window Shades

Thermal shades, shutters, or heavy curtains can reduce heat loss through windows at night by up to 80 percent. Many types of night insulation can be hand-made. Others can be purchased and professionally installed. A snug fit and tight edges all around are important for high

A Super-insulated Wall

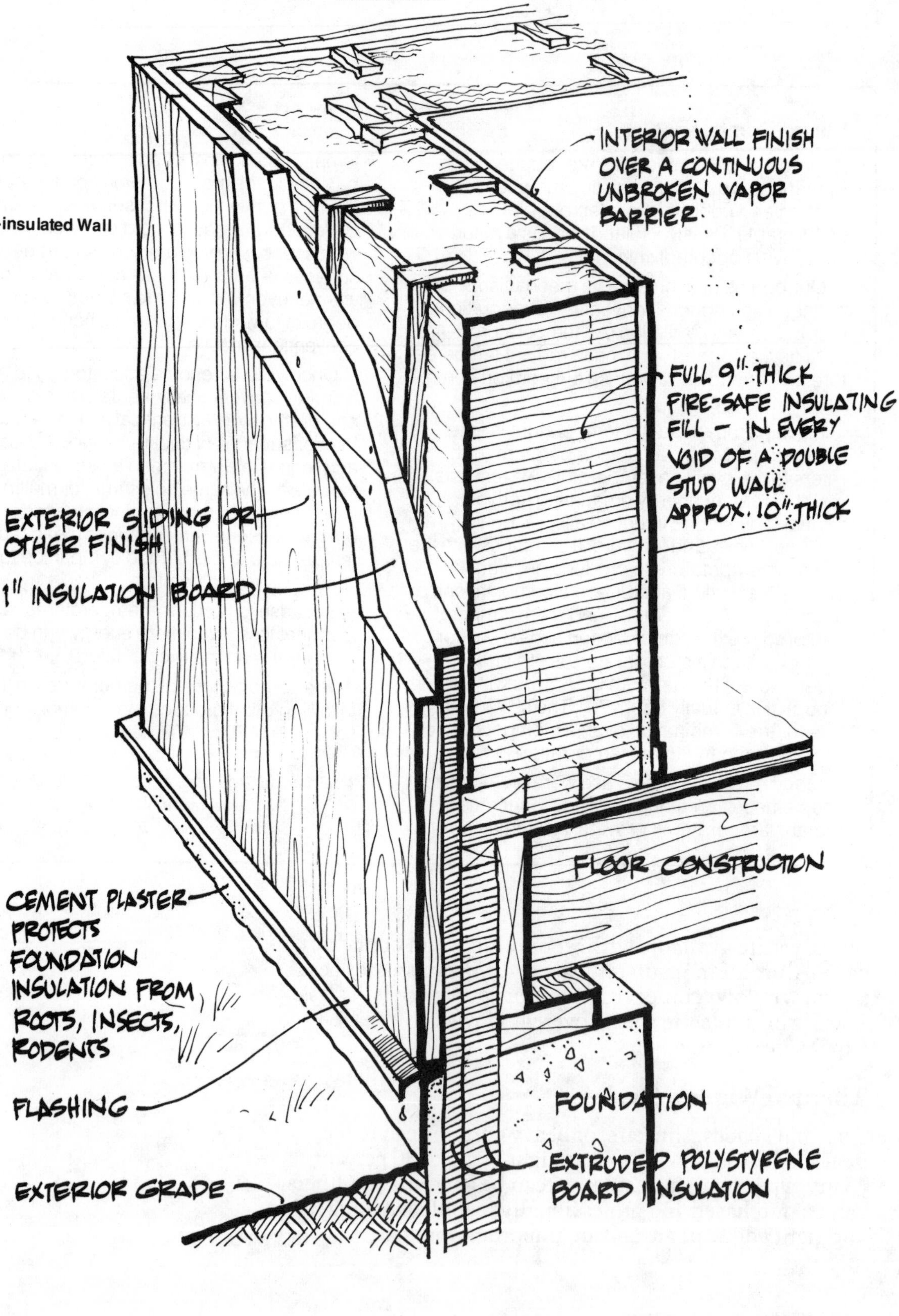

INTERIOR WALL FINISH OVER A CONTINUOUS UNBROKEN VAPOR BARRIER

FULL 9" THICK FIRE-SAFE INSULATING FILL — IN EVERY VOID OF A DOUBLE STUD WALL APPROX. 10" THICK

EXTERIOR SIDING OR OTHER FINISH

1" INSULATION BOARD

FLOOR CONSTRUCTION

CEMENT PLASTER PROTECTS FOUNDATION INSULATION FROM ROOTS, INSECTS, RODENTS

FLASHING

FOUNDATION

EXTERIOR GRADE

EXTRUDED POLYSTYRENE BOARD INSULATION

effectiveness. Combining night insulation with double or triple glazing will be the greatest deterrent to nighttime heat loss through windows.

Shades and Awnings

To stay cool during the summer, keep the sun out of the house during the day. Inexpensive interior shades from a hardware store are least effective but work well for the small investment. Exterior awnings are becoming popular again and do a good job of shading. Deciduous trees provide ideal summer shading and shed their leaves to allow winter sun in. Summer shading by whatever means is a "natural" first step in reducing cooling costs.

Taking Steps

Take time to get sound advice and to implement energy conservation measures properly. Although procedures are often simple, inadequate or faulty installations can reduce insulating effectiveness or even damage the house. But, without a doubt, energy conservation will pay you well for your effort in fuel savings.

Solar Retrofit

Suppose you already own a home and want to keep it, but are concerned about rising energy costs. Solar heating and cooling are not out of your reach. Solar retrofitting, or adding solar features to existing homes, is one of the most exciting challenges in the field.

There are a lot of existing homes and all of them use energy, often more than necessary. For many reasons— economic, historic, aesthetic or purely sentimental—we don't want to just discard older homes or other valuable buildings. Solar renovating or retrofitting is a viable option to consider.

All the passive solar choices we've talked about come in "retrofit sizes" too. Take time to compare options before choosing. Some solar retrofits will suit your existing structure better than others. But retrofit possibilities are not necessarily limited either. If you use your imagination and then carefully check out the most practical options, the results will be very satisfying. Some attractive and efficient solar retrofits are featured in the eight page photo spread on pp. 119–126.

Do button up first. Older homes, even those built five years ago, rarely have enough insulation or are tight enough to maximize the performance of any solar retrofit. Research for your solar retrofit starts very simply. Find south. Every home has a south-facing wall or corner at worst. If the compass does not yield a perfect long wall facing due south, don't give up. Orientation can be up to 30° off either to the east or west of south and still be effective for solar collection.

If, when looking in a southerly direction, you don't find a high rise, you're in luck. Any obstruction which casts a shadow on the house in winter can reduce solar collection unless it can be removed or like a deciduous tree loses its leaves when you will need the winter sun. Summer shading from deciduous trees is a cooling advantage too. Solar orientation and shading factors are only the first steps in evaluating site suitability.

If the site checks out so far, think of the five passive solar options. Which ones make sense for you? It will depend upon the design of your home, its position on your lot, and the dollar investment you are prepared to make. For example, a solar room will make sense if you like gardening or want extra living space in addition to solar heat. If you want solar heat and privacy is needed because your south side faces a street, a solar chimney for a frame house or a solar wall for a masonry or frame house makes the most sense. Adding more south facing windows is a simple, efficient solution in many cases. If the only available south-facing surface is a roof, a solar attic may be a natural choice.

Cost is an important consideration in passive solar retrofits. (When building with a new passive home design, extra cost may be minimal when compared with a new conventional home.) A solar retrofit, like any remodeling work, will cost money, but compared to what? If you convert the home you have to solar instead of building a new one you could save a lot of money by comparison and substantially reduce heating and cooling cost as well.

Passive solar retrofits come in many "sizes," as well as "cost variations." A greenhouse solar room, for example, can vary from light lean-to framing covered with plastic glazing to one which is custom-built, with triple-glazed windows, built-in shutters, and water-wall heat storage. Both are appropriate for some uses and both function well. Cost is the variable, so you must decide what you want to spend.

To help you make really informed retrofit decisions, each of the five passive heating options will be critically considered. However, one more bit of information you need is energy flow in and out of buildings. Heat and light flow through windows. Heat travels through walls also. The important point is that there are two primary flows of heat in and out of the house. One is solar radiation inward; the other is heat escaping from your house in cold weather and seeping in during hot. Both vary considerably depending on the time of day and the season of the year.

Adding transparent glazing to a house is the basic strategy in solar retrofitting. It is a way of taking advantage of needed energy flows in and reducing unwanted energy flows out. Solar retrofit strategy assumes that conservation measures have been taken first. When a second glazing layer is added to an existing window, A, it greatly reduces heat loss, but reduces solar heat gain only slightly. When glazing is added to the prepared wall surface of a house, B, it transforms the wall into a solar chimney collector. Adding a layer of glazing to an uninsulated masonry wall, C, significantly reduces heat loss and, in fact, produces considerable solar heat gain. Finally, a solar room results, D, when the space between the glazing and the wall is greatly enlarged. The space, in a sense, absorbs the shock of outdoor weather extremes, tempering their effect on the house while also providing solar heat.

The Ins and Outs of Energy Flow

If you understand how energy flows through windows and walls, you can more easily select the most suitable passive design for your house. There are two primary flows of heat in a house. One is solar radiation inward; the other is heat that escapes from your house in cold weather and seeps into your house during hot weather. Both types of flow vary considerably in amount depending on the time of day and the season of the year. Keep energy flow in mind as we look at the five basic passive solar options in a simple way.

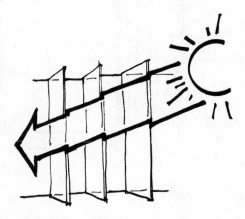

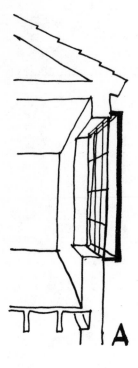

All are variations of adding extra glass to let solar heat in and trap it to prevent heat losses back out.

Transparent glazing can be added to south sides of houses in a number of ways to affect the amount of both solar radiation that enters into, and heat that escapes from, the house.

A. When glazing is added over an existing window, it greatly reduces heat loss but reduces solar gain only slightly.

B. When glazing is added to the prepared wall surface of a house, it transforms the wall into a solar chimney collector.

C. Adding a layer of glazing over an uninsulated masonry wall significantly reduces heat loss from that wall, and, in fact, produces considerable solar heat gain through the wall.

D. Finally, a solar room results when the space between the glazing and the wall is greatly enlarged. This space, in a sense, absorbs the shock of outdoor weather extremes, tempering their effect on the house while also providing solar heat.

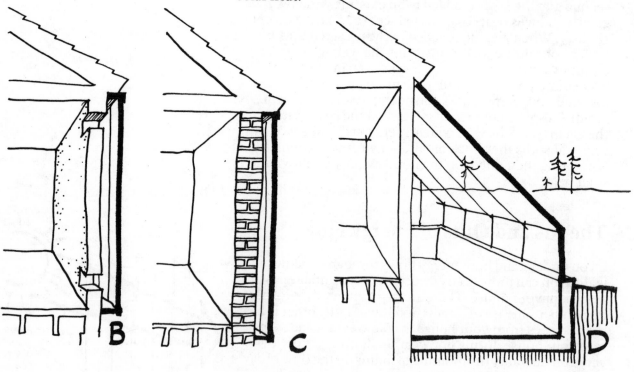

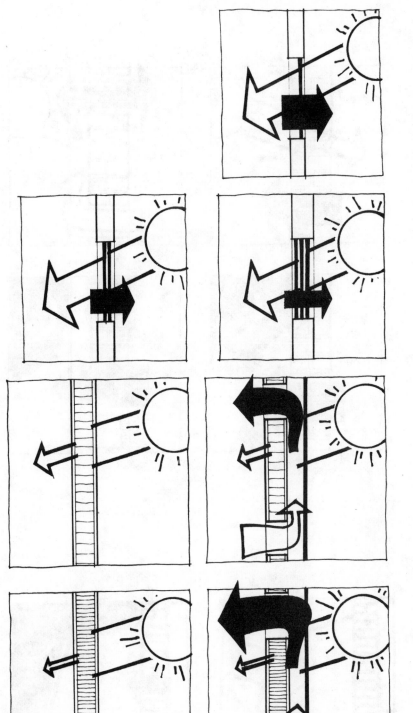

A. Solar Windows

Improving the Energy Performance of Existing Windows. Levels both of solar radiation "in" and of heat loss "out" are high through a single layer of glass. When the sun is shining, the house heats up quickly. Yet, when the temperature drops, heat loss increases quickly. A second layer of glass reduces solar gains by about 18%, but reduces heat loss by about 50%. A third layer of glass reduces solar heat gain by another 18%, but heat loss is reduced by an additional one-third. Therefore, a second, and even a third, layer of glazing is often cost effective. Movable insulation (to be discussed in Chapter 3) is most effective in reducing heat loss.

The same principles apply if you convert south-facing walls into windows, perhaps the simplest solar retrofit. Added glass area allows more heat in, but be sure to take steps to reduce heat loss. Be sure also to take steps to soak up the extra heat to keep the house comfortable during the day and to make the extra heat available at night. Make provisions for shading the glass during the summer.

B. Solar Chimneys

Converting Heat-losing Walls into Energy Producers. A poorly-insulated wall allows small amounts of solar heat gain. A well-insulated wall allows little or no solar gain. Whereas the poorly-insulated wall has huge heat losses, the well-insulated wall loses very little. When the outdoor temperature drops, heat loss through the walls increases quickly but not as quickly as it does through windows.

When the walls are covered by sheets of glass, and thereby are converted to solar chimney collectors, they increase considerably the amount of solar energy they provide to the house. They take a short time before they heat up and start producing heat, so they do not provide energy quite as quickly as windows do. Nor do they provide quite as much heat as windows do. However, the heat loss from the house through the walls is substantially less than through the windows (unless, of course, the windows are covered with insulation at night). The net result is that more energy is gained through solar chimneys than is lost. And, if properly constructed, solar chimneys can produce more energy than windows can.

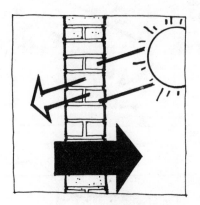

C. Solar Walls

Brick 'n Mortar 'n Solar. Brick, stone, adobe, and concrete walls have high rates of heat loss, even if they are thick. If they face east, west, or north, insulate them, preferably with the insulation on the outside. But if they face south, cover them with sheets of glass or durable plastic and capture the sun's rays.

Much of the sun's heat is absorbed by the wall, delaying the time when the house receives the heat. Also, because the sun-warmed wall loses heat back to the outdoors, the net energy gain is not as great as it is through windows. But the solar heat enters the house slowly and over a long period of time, making overheating much less of a problem than with solar windows, and keeping the house warm well into the night.

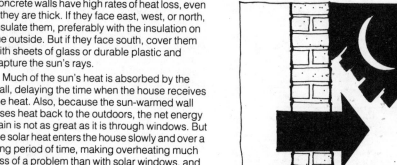

D. Solar Rooms

Going One Step Further—Solar Rooms. Vertical glazing offers only a small, dead air space over an exterior wall surface. If the glazing is installed instead in a lean-to fashion, the air space can become large and can be called a solar room. The heat loss from the house is no longer to the outdoors, but rather it is to this large air space, which is nearly always warmer than the outdoors. This makes the rate and amount of heat loss from the house much lower.

If the wall of the building is wood framed, a solar room is likely to experience wide temperature fluctuations. If the wall is of solid masonry, then the fluctuations will be much smaller. The thermal mass of masonry or earthen floors reduces temperature fluctuations, too.

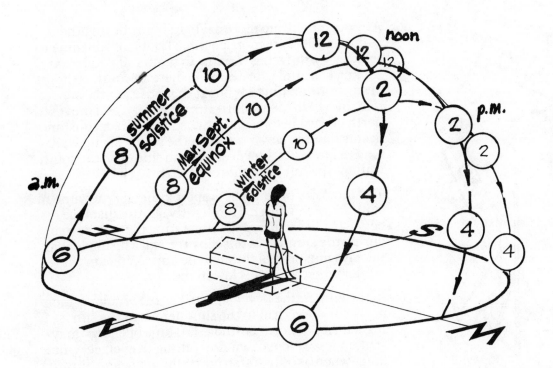

Instructions: Stand at the line of your proposed solar wall, face south, look at the numbered suns in the sky, wait there for 12 months until all 17 have appeared, then continue reading this book.

Solar Position

We all realize that the sun doesn't stay in one part of the sky all day and that its path varies from season to season and from state to state. Fortunately, its movements are completely predictable, widely published, and easy to understand. No guesswork is involved: from the sun chart for your latitude (see Appendix 1) you can find quite easily where the sun will be at any hour, at any season; and, from that information you can see how and where solar installations (and summer sunshades) must be placed in order to respond to the sun where you live.

Here's a nice surprise: the quantity of solar energy that penetrates a south-facing window on an average sunny day in the *winter* is greater than that through the same window on an average sunny day in the *summer*. Here are the reasons:

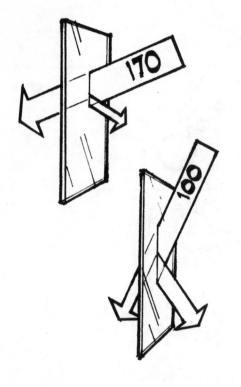

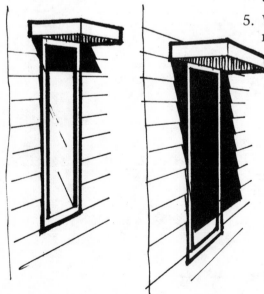

1. Although there are more daylight hours during the summer, there are more possible hours for sunshine to strike a south-facing window in winter. If you live at 35° north latitude, for example, there are fourteen hours of sunshine on June 21. But at that position the sun remains north of east until after 8:30 A.M. and moves to north of west before 3:30 P.M., so that direct sunshine occurs for only *seven hours* on the south-facing wall. On December 21, however, the sun shines on the south wall for the full *ten hours* of daylight.

2. The intensity of sunlight is approximately the same in summer as in winter. The slightly shorter distance between the earth and the sun during the winter than during the summer is offset by the extra distance that the rays must travel through the atmosphere in the winter when the sun is low in the sky.

3. In the winter, the lower sun strikes the windows more nearly head-on than in the summer when the sun is higher. At 35° north latitude, 170 Btus of energy may strike a square foot of window during an average winter hour, whereas only 100 strike on the average during the summer.

4. In winter, more sunlight passes through glass by hitting the window head-on. But in the summer, the high-angle rays tend to reflect off the glass.

5. With proper shading, windows can be shielded from most of the direct summer radiation.

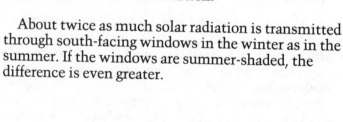

    About twice as much solar radiation is transmitted through south-facing windows in the winter as in the summer. If the windows are summer-shaded, the difference is even greater.

In passive systems, tilted surfaces such as roofs are used less often than vertical surfaces. With reflective surfaces such as snow on the ground, a south-facing vertical surface actually receives more energy during the middle of the winter than a south-facing tilted one. Therefore, during the primary heating months there is little advantage to using tilted rather than vertical, south-facing surfaces. In fact, for more northern latitudes, the difference is insignificant.

Tilted glazing, whether in collectors or skylights, tends to be more costly to build and more prone to leakage. It is also harder to shade and, if left unshaded, can more easily overheat the house in the summer than vertical glazing can. Roofs are less likely than walls to be shaded by trees or buildings during the winter, and they have large surfaces for collecting solar energy. Unfortunately, they are difficult to cover with insulation at night to reduce heat loss.

The Site

If your site does not have proper solar exposure because it slopes sharply north or is shaded darkly by evergreens or large buildings, a house designed for the site will have little chance of being solar heated. Here's what to watch for:

1. Lot Orientation

South-facing houses assure lower energy consumption, during both summer and winter. This does not mean that houses have to face rigidly southward. A designer who understands passive solar principles can devise dozens of practical solutions. The site or lot itself does not have to face south, as long as the building itself is oriented southward. A lot that slopes sharply north is, of course,

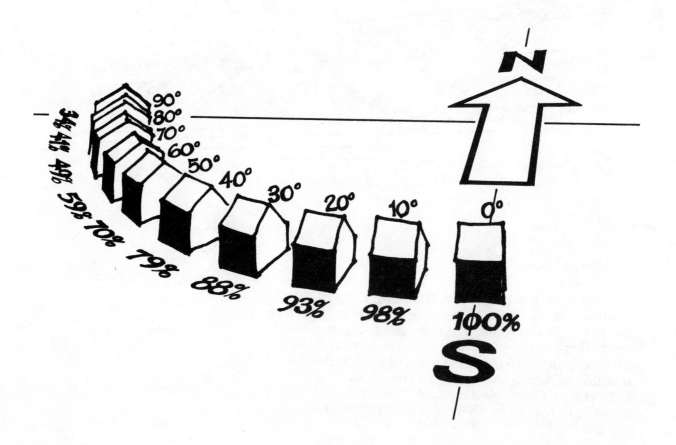

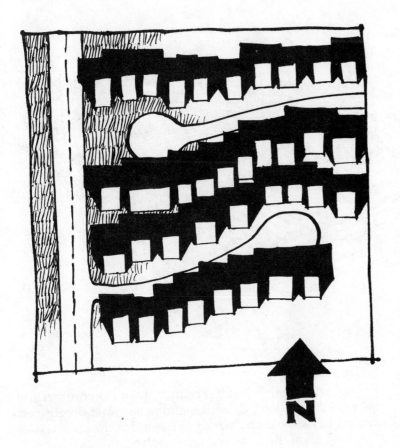

very difficult to work with, and south-sloping lots are preferred. Once land developers understand passive principles, they can plan for solar subdivisions with the cost approximately the same as for conventional subdivisions.

2. Setback Flexibility and Minimum Lot Size

Deep house lots which have narrow street frontage, reduce the surface area of summer heat-producing asphalt streets. Higher housing densities can reduce travel distances and times and subsequent energy use. Flexible zoning laws can permit houses to be located near the edges of their lots, thus minimizing the potential of shading from adjacent neighbors. Long-term, shade-free rights to the sun are necessary to guarantee adequate sunlight for the life of the house. Solar rights are slowly being acknowledged as legal precedents build up in the courts.

3. Landscaping

Proper landscaping can offer beauty as well as comfort and energy savings, both in winter and in summer. Evergreens can greatly slow arctic winds. For most of the country, these winds come from the west, north, and northwest. Large deciduous trees appropriate to your region can provide shade and summer cooling. They are most effective on the east, west, and south sides of the house. Most, but not all, deciduous trees shed their leaves in the winter to let the warm sun in. Well-shaded and landscaped paving will often encourage people to walk or bicycle rather than ride in an energy-consuming car. Glaring, unshaded asphalt creates desert-like conditions, placing a higher air conditioning load on buildings. Pavings that are porous to rain and that do not absorb heat have a much less severe effect.

Landscape design that encourages home vegetable gardening saves energy in many ways. For each calorie of food produced by agriculture, ten calories of manufactured energy are expended. Home gardens do far better, offering not only more nutritional food, lower food bills, and richer soil, but also a more appropriate use for all the plant cuttings and food wastes so often discarded.

4. Street Widths

Narrow streets save valuable land and can be shaded more easily than wide ones. They are more pleasant than wide multilaned streets and are safer for bicyclists, pedestrians, and motorists. They reduce the heat load on people using them, and they also reduce traffic speeds. Parking bays, rather than on-street parking, can promote shading both over the bays and over the narrower streets. Pedestrian walks and bicycle paths are far more readily integrated into such plans.

5. Length/Width Ratios

In the northern part of the country, south sides of houses receive nearly twice as much radiation in the winter as in the summer. This is because the sun is lower in the sky during the winter. In the summer, the sun is high in the sky, and the sun does not shine directly on south walls for a very long period of time. Houses in the south gain even more on south sides in the winter than in the summer.

East and west walls receive 2½ times more sunshine in the summer than in the winter. Therefore, the best houses are longer in the east/west direction, and the poorest are longer in the north/south direction.

A square house is neither the best nor the worst. (Remember, however, that a square building is often the most efficient in terms of layout and economy of materials.) A poorly shaped house can be improved by covering the south wall with windows and other passive systems and minimizing windows facing other directions. If you pitch your roof south at a slope of 45° or so, you can add active solar collectors and/or photovoltaic (solar electric) cells. You may not want to do so now, but someday when energy prices have tripled, and then tripled again, everyone will envy your farsightedness. After all, this year's $150 heating bill of your cozy passive solar home will at that point be over $1,000 while everyone else's will be $5,000 to $10,000!

6. Natural Daylighting

Do not underestimate the bonus of natural daylighting, which passive solar designs can provide. In some big buildings, solar glazing may save more energy and money by reducing electric light bills than it saves by reducing fuel bills, and lighting engineers feel that properly located lighting from sidewalls can be two to three times as effective as artificial overhead lighting. For houses, the extra light from solar windows and solar rooms can add immeasurable pleasure and a living experience far surpassing any you've had before.

PART 2

Passive Solar Heating

Horace Walpole, sitting by the large solar window in his library at Strawberry Hill. (Wash drawing, 1756. The Bettmann Archive.)

Chapter 3

Solar Windows

Sunlight falling gently through windows is by far the most common way for solar energy to heat our country's 70,000,000 buildings. But loosely fitting, single-glazed windows usually lose more heat than they contribute in the form of solar heat gain. On the other hand, a properly-designed south window, with the addition of a reflective surface on the ground (such as snow, a pond, or an aluminized mirror) and with an insulating cover at night, can supply up to twice as much solar energy to a building as a good "active" solar collector of the same surface area.

Proper design criteria include the following:

The right **timing**. Sunlight must enter a house at only the right time of the year and the right times of the day. This simply means careful design in response to known solar geometry and climate.

The right **amount** of solar heat. If you desire fairly stable indoor temperatures, this must be engineered. If you desire a fairly wide range of temperatures, great—but just don't assume that you do, or that other occupants or your friends will find it comfortable, and then have to later excuse a poorly engineered system when it gets too hot. Too much glass and too little mass is a common but unnecessary error in properly well-insulated passive solar houses.

The right **type** of glazing. Clear glass is both attractive and efficient and has its place, but so have clear and translucent materials, such as diffusing glass, plastic films, fiberglass-based glazings, and acrylics. Reflective glazings can reduce unwanted solar gains. (See Appendix 6 for more help on selecting the right glazing material.)

In one sense, of course, most of the world's buildings are already principally solar heated. Think about the huge, year-round heating job the sun does, using land and sea as thermal mass, to keep the whole world between −50°F and +100°F when, without the sun, we'd be at −473°F all the time. That's *most of all of the heating done on earth*, and we'd do well to remember it when we hear talk of alternatives to solar energy use or that solar can't really do the job.

41

Windows and Solar Collectors Compared

Properly designed solar collectors supply between 50,000 and 85,000 Btus per square foot of surface area per heating season in a climate where the sun shines half the time. (This is equivalent to the energy from ½ to 1 gallon of home heating oil, or from 15 to 25 kwh of electricity, or from 100 to 140 cubic feet of gas.) Solar gain through a square foot of south-facing, double glass in the same climate is about 140,000 Btus. Conduction heat loss through that square foot (ignoring air infiltration for the moment) is about 70,000 Btus in a 5,000 degree day climate. The net contribution to the building, then, is 70,000 Btus (140,00 solar gain less 70,000 heat loss). Therefore, in a climate like that of St. Louis, ordinary double-glazed south-facing windows can produce about the same amount of heat per square foot as solar collectors. Reflectors will boost heat production of both designs. Movable insulation and/or triple-glazing can dramatically reduce heat loss from windows, greatly boosting their net energy input to the house.

Solar Gain

Appendix 2 provides month-by-month, hour-by-hour clear-day sunlight (or "insolation") data for vertical, south-facing surfaces for six different northern latitudes. Together with Appendix 3 (U.S. sunshine maps for each month), it can be used to determine the approximate amount of solar radiation likely to come through south-facing glass anywhere in the United States at anytime.

For example, from Appendix 2, the total clear-day solar radiation on a south-facing, vertical surface in January at 40° north latitude (Philadelphia, Kansas City) is 1726 Btus per square foot. In Kansas City, the "mean (average) percentage of possible sunshine" in January is 50 percent. (See the U.S. sunshine map for January.) Approximately 82 percent of the sunshine that hits a layer of ordinary glass during the day actually gets through it. Therefore, the average total amount of solar radiation penetrating one layer of vertical, south-facing, double-strength glass in Kansas City during the month of January is approximately

(31 days per month) × (1726 Btus per square foot per day)
   × (50 percent possible sunshine)
   × (82 percent transmittance)
   = 22,000 Btus per square foot per month.

Over the course of a normal heating season in a 50-percent-possible-sunshine climate, the total solar gain will be between 130,000 and 190,000 Btus per square foot. (A more accurate number may be obtained by doing the calculations on a month-by-month basis for each month of the heating season for a particular location, taking into account the heating needs of the particular house.)

If a second layer of clear glass is added to the first, about 82 percent of the light that penetrates the first layer will penetrate the second. Converting the first example, then, 18,000 Btus (0.82 x 22,000) are transmitted by double glass compared with 22,000 by single. But remember heat losses are reduced by 50 percent when you do this!

These monthly solar gains can be roughly compared with the monthly heating demands of the house to determine the percent of the heat supplied by the sun. When solar provides less than 40 percent of the heat, the above analysis is relatively accurate for preliminry design purposes. However, a more detailed and rigorous analysis is required when the solar windows are large enough to be supplying more than 40 percent of the heating load.

In cold climates, 300 square feet of direct gain will supply roughly half the heat for a well-insulated, 1500 square foot house. Half as much area is needed in a mild climate.

Minimum heat loss. Keep heat losses back out through the glazing as low as is practical. Use several layers of glazing according to the material and climate. Cold climates also warrant movable insulation at night.

Control of glare and fading. Some people simply do not like working in direct sunlight. In fact, many people prefer the softer north light. Southern exposure means low fuel bills, but it also means window glare and squinting. Too much glass can also mean loss of privacy. Overhead light (such as from a skylight) is often a good compromise, offering solar gain with the least glare. In colder climates, however, this can mean added heat loss at night. Make sure the overhead glass is shaded during the summer!

Thermal Mass

The sun does not shine twenty-four hours a day, and thus, unlike a furnace, it is not waiting on call to supply us with heat whenever we need it. Therefore, when we depend on the sun for heat, we must do as nature does—store the sun's energy when it is shining for use when it is not. Nature stores the sun's energy a number of ways. Plants use photosynthesis during the day, and then they rest at night. Lakes become heated during the day and maintain relatively constant temperatures day and night. For hundreds of years people have been growing and harvesting food during the summer and storing it for use during the winter. Indians of the American Southwest have for centuries used thick adobe walls that act like big thermal sponges to soak up large amounts of sunlight. As their exterior surfaces warm up during the day, the heat slowly moves throughout the adobe, protecting the interior from overheating. At night, the walls cool off, allowing the adobe to soak up heat again the next day, thus keeping the houses cool.

In contrast, massive central masonry chimneys of New England colonial houses absorb any excess *interior* daytime warmth. The stored heat helps keep the houses warm well through the night.

When massive materials are located inside houses where the sun can strike them directly, they combine the

Rules of Thumb for Thermal Mass

If sunlight strikes directly on the mass (such as a brick floor), each square foot of a window needs roughly 2 cubic feet of concrete, brick, or stone to prevent overheating and to provide heat at night. If sunlight does not strike the mass, but heats the air that in turn heats the mass, four times as much mass is required.

The ability of a material to store heat is rated by its "specific heat," meaning the number of Btus required to raise 1 pound of the material 1°F in temperature. Water, which is the standard by which other materials are rated, has a specific heat of 1.0, which means that 1 Btu is required to raise 1 pound of water 1°. The pound of water, in turn, releases 1 Btu when it drops 1°.

The specific heat of materials that might be considered for use in the construction of buildings are listed below. The second column of numbers in the table shows the densities of the materials in relation to each other. The material's heat capacity per cubic foot (listed in the third column) was obtained by muliplying its specific heat by its

Specific Heats, Densities, and Heat Capacities of Common Materials

Material	Specific Heat (Btus stored per pound per degree change of temperature)	Density (pounds per cubic foot)	Heat Capacity (Btus stored per cubic foot per degree change of temperature)
Air (75°F)	0.24	0.075	0.018
Sand	0.191	94.6	18.1
White pine	0.67	27.0	18.1
Gypsum	0.26	78.0	20.3
Adobe	0.24	106.0	25.0
White oak	0.57	47.0	26.8
Concrete	0.20	140.0	28.0
Brick	0.20	140.0	28.0
Heavy stone	0.21	180.0	38.0
Water	1.00	62.5	62.5

density. Note that the density of water is least among the materials listed but that its heat capacity per cubic foot is still highest because of its high specific heat. The low specific heat of concrete (0.2, or 1/5 that of water) is partially compensated by its heavy weight and it stores considerable heat (28 Btus per cubic foot for concrete, or about one-half that of 62.5 for water). Except for water, the best readily available materials are concrete, brick, and stone.

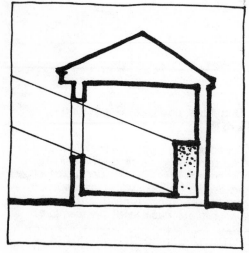

benefits of Southwest adobe and New England chimneys. The resulting "thermal mass" tempers the overheating effects of sunlight from large windows and absorbs excess energy for later use.

Generally, the more thermal mass the better. But if its *too* thick, heat may not get through. The more directly the sun strikes the mass, the less the house temperature will fluctuate. Unfortunately, thermal mass, such as brick walls, concrete floors, or water storage tubes, are often expensive and/or unsuitable to the homeowner. Thus, moderately-sized solar windows, which require limited amounts of thermal mass, are often the best solution. Solar walls and/or solar rooms can supplement the solar windows to achieve the lowest possible fuel bills and the highest possible levels of comfort.

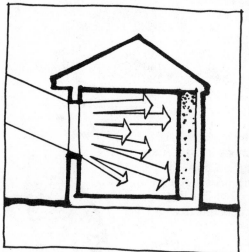

An unheated, lightweight house, such as a wood-framed one, drops in temperature relatively quickly even if it is well-insulated. A heavy, massive, well-insulated structure built of concrete, brick, or stone maintains its temperature longer. To be most effective, the heavy materials should be on the inside of the insulating envelope of the house. When left unheated, a house that is well insulated and also buried into the side of a hill cools off very slowly and eventually reaches a temperature close to that of the soil. Although earth is a good means of sheltering your house from the extremes of weather, soil is a poor insulator and will draw heat out of the building endlessly if you don't insulate well.

If you prefer to close draperies to keep the sun out, or if you insist on wall-to-wall carpeting or big rugs, solar windows might not make sense for you. Alternatively, reconsider how you desire to furnish your house. The warmth of brick floors, walls, and fireplaces and the sensation of light and heat coming through windows can be exhilirating, possibly more so than wall-to-wall synthetic fabrics.

Clear glass allows the bright rays of the sun to shine directly on specific surfaces in the room, leaving others in shadow. Translucent glazing, on the other hand, diffuses light and distributes it more widely, assuring more even heat distribution to many interior surfaces at the same time. This results in more even temperatures and greater heat absorption and storage throughout.

The temperature swings of thermal mass placed in direct sunlight will be about twice the temperature swings of the room itself. Mass shaded from the sun inside the room (such as in north walls) will fluctuate in temperature about half as much as the room. Thus, solar radiated mass stores four times more energy than the shaded mass.

Too little heat storage will allow wide temperature swings and permit overheating, which in turn wastes heat

Concrete Floors

Concrete floors are commonly used for storing heat from solar windows. Consider this oversimplified case: A 20 by 40 foot house has a concrete floor 8 inches thick (530 cubic feet). By late afternoon the slab has been solar heated by 150 square feet of window to an average of 75°F. During the night, the outdoor temperature averages 25° and the indoor air averages 65°. A well-insulated house may lose heat at a rate of about 200 Btus per hour for each degree of temperature difference (called Delta T, or ΔT) between the outdoors and indoors. The temperature of the slab drops as it loses heat to the house.

The heat lost from the house is the product of the total heat loss rate, the time, and the average temperature difference between indoors and outdoors. In this case, the heat loss during the 15 hour winter night is

$$(15 \text{ hours}) \times (200 \text{ Btus per hour per °F}) \times (65° - 25°) = 120,000 \text{ Btus.}$$

With a heat capacity of about 28 Btus per cubic foot per degree of temperature change, the 530 cubic foot concrete slab stores roughly 15,000 Btus for each degree rise in temperature. for each degree drop in temperature, the slab releases the same 15,000 Btus. If the floor drops 8 degrees, from 75°F to 67°F, it will release just enough heat, 120,000 Btus, to replace the heat lost by the house during the night.

When you calculate mass floor areas, realize that the mass must be left exposed in order to work. Although concrete floors perform well, even the best designed floors for solar exposure very often get covered or shaded by rugs or furniture.

due to greater heat loss from the house, (especially if you open windows to vent that extra heat). Conversely, more mass increases both comfort and the efficiency of the passive system.

The effectiveness of mass also depends on its thickness. The deeper parts of thick walls and floors are insulated by the surface layers and do not store as much heat. Therefore, 100 square feet of 8-inch-thick wall is more effective than 50 square feet of 16-inch-thick wall, even though they both weigh the same.

Provide for thermal mass in the simplest way possible, otherwise it can be costly and can complicate construction. When used wisely, on-site locally available building materials (gravel, stone, etc.) can be the best kinds of thermal mass. Their use requires less energy than it takes to make and transport brick and concrete.

Movable Insulation

Glass loses heat up to 30 times faster than well-insulated walls, so the nighttime insulation of glass in winter climates is very important. So is the use of double glazing, which has only half the loss rate of single glass. If the double glazing faces south, it gains more heat than it loses during the winter, virtually anywhere in the country.

In climates of more than 5,000 degree days, the extra cost of triple glazing is usually justified by the energy savings. However, more than three-layered glass seldom is, since each layer of glass also blocks fifteen to twenty percent of the solar energy that passes through the preceding layer. Multilayered, nonglass glazing systems of high transmittance (up to 97 percent), such as Teflon™, can often use four or five layers effectively. The reason for this is that they are so clear that an additional layer

---

Energy Savings from Movable Insulation

To determine the annual energy savings using movable insulation, first find the difference between the U-value of the window as it is and the U-value of the window using movable insulation. Then multiply this difference times the number of degree days where you live times 24 hours per day.

For example, suppose that an insulating panel with a heat flow resistance of R-10 is being considered for windows in Minneapolis with two layers of glass. Assume that the insulation will be in place an average of 12 hours per day. The U-value of the glass is 0.55 Btus per square foot per hour per degree (from Appendix 4). The U-value for the insulated window system is 0.24. The difference between the two is 0.31 (0.55 minus 0.24). Minneapolis averages 8382 degree days per year (from Appendix 5). Therefore,

Annual energy savings
= (0.31 Btus per square foot per °F)
   × (8382 degree days per year)
   × (24 hours per day)
= 62,362 Btus per square foot per year

For a 10-square-foot window, the savings is roughly equivalent to the heating energy obtained from 180 kwh of electric resistance heating ($7–$12 at most electric rates), from 10 gallons of oil burned in most furnaces, or from 10 square feet of an active solar collector of average design. A tight-fitting shutter also reduces heat loss due to air leakage around the window frame, making the above savings a conservative estimate.

reduces heat loss significantly, yet blocks very little of the incoming sun. These thin plastics are not commonly used in home construction; however, for existing homes, thin film plastics are frequently used in place of glass storm windows. Companies are developing products that will make thin filmed plastic windows easier to use for both new and existing homes.

But the most direct options for preventing unwanted heat loss through solar windows are:

- sheets of rigid insulation manually inserted at night and removed in the morning;

- framed and hinged insulation panels;

- roller-like shade devices of one or more sheets of aluminized mylar, sometimes in combination with cloth and other materials;

- sun-powered louvres, such as Skylid™, which automatically open when the sun shines and close when it doesn't; and

● mechanically-powered systems, such as Beadwall™,
which use blowers to fill the air space between two
layers of glazing with insulating beads at night.

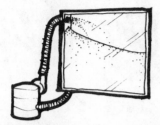

The insulating values of good movable insulating
devices range in heat flow resistance from R-4 to R-10.
During the day when the sun is shining, windows are net
energy producers. But since outdoor temperatures are
much lower at night, up to three quarters of a window's
24-hour heat loss can be prevented by the proper use of
these devices.

A window loses heat to the out-of-doors in proportion to
the temperature of the air space between the window and
the insulation provided. A loose-fitting insulating shutter
will allow room air into that space and diminish the
insulating effect. Therefore, a snug fit and sealed edges are
important.

A few cautioning words: Sun shining on an ordinary
window covered on the inside by a tight insulating shade
can create enough thermal stress to break the glass. A
white or highly-reflective surface facing the glass is the
best solution, but not foolproof. Also, moisture can
condense at night on the cold window glass facing the
insulation, causing deterioration of the wooden frames.
Tight-fitting insulation is the best solution for preventing
excessive condensation. Otherwise, provisions for
collecting and draining the condensation may be
necessary.

Also, remember to conform to all codes; don't use
insulation materials that are flammable or in other ways
hazardous without protecting them properly.

Chapter 4

Solar Chimneys

A solar chimney is an air-heating solar collector that runs automatically, on sun power alone. Of all the passive heating systems, it loses the least heat when the sun is not shining. Except for solar windows, solar chimneys (also called convective loops) are the most common solar heating systems in the world.

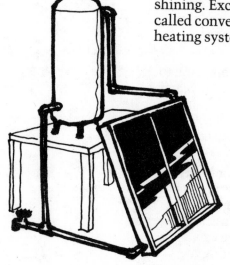

Variations on the design are used to heat water for domestic purposes. Hundreds of thousands of pumpless, "thermosiphoning" (heat convecting) solar water heaters have been used for decades. In fact, convective loop water heaters were patented in 1909. By 1918, 4,000 such water heaters were in operation in Southern California.

Solar chimney wall-mounted collectors can complement south-facing windows in supplying additional solar heat directly to both new and existing houses. In conventional wood-framed houses, up to 25 percent of the heat can be supplied by combining solar windows and solar chimneys without supplemental thermal storage. A combined system of

roughly 200 square feet can achieve the 25 percent figure for a well-insulated 1500-square-foot house in a cold climate. Half as much area is needed in a mild climate.

Basic System Design

A solar chimney wall collector is similar to a flat-plate collector used for active systems. A layer or two of glass or plastic covers a black absorber. Air may flow in a channel either in front of or behind the absorber, depending on the design. The air may also flow *through* the absorber if it is perforated. The collector is backed by insulation.

In the figure on the next page, the collector is mounted on, or made a part of, the insulated wall. Openings at the bottom and top of the wall permit cooler air from the house to enter the hot collector at the base of the wall, to rise as the sun heats it, and to vent back into the building near the ceiling.

The slow-moving collector air must be able to come in contact with as much of the absorber's surface area as possible without being slowed down too much. In fact, the amount of heat transferred from the absorber to the flowing air is in direct proportion to the heat-transfer capabilities of the absorber and the speed of the air flow by or through it.

Up to six layers of expanded metal lath is used in some absorbers. In these, the air rises in front of the lath, passes through it, and leaves the collector through a channel behind the lath. Flat or corrugated metal is also used, but it does not transfer heat as well. However, the air flow channel in this case need not be as deep. The metal should

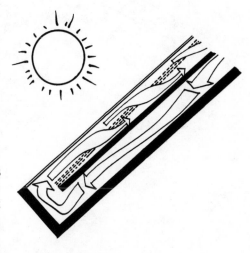

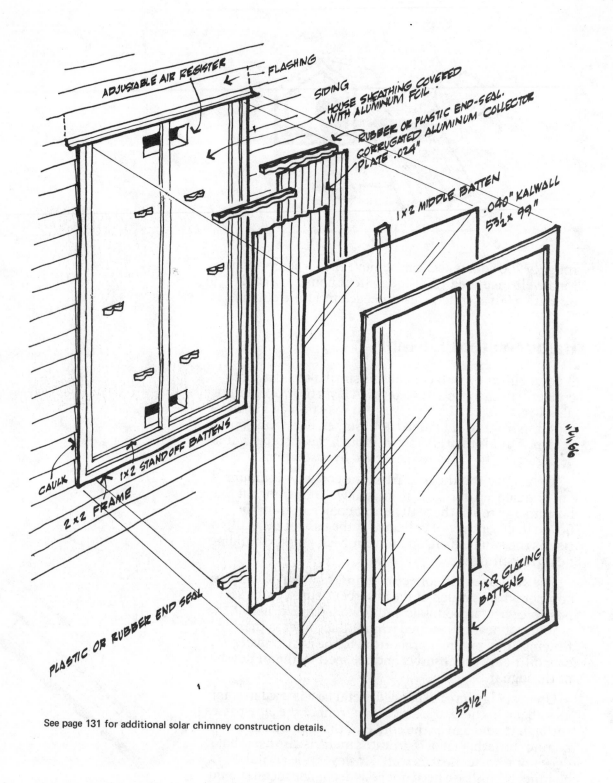

ADJUSTABLE AIR REGISTER

FLASHING

HOUSE SHEATHING COVERED WITH ALUMINUM FOIL

SIDING

RUBBER OR PLASTIC END-SEAL

CORRUGATED ALUMINUM COLLECTOR PLATE .024"

1×2 MIDDLE BATTEN

.040" KALWALL 53½ × 99"

99½"

1×2 STANDOFF BATTENS

CAULK

2×2 FRAME

1×2 GLAZING BATTENS

PLASTIC OR RUBBER END SEAL

53½"

See page 131 for additional solar chimney construction details.

be placed in the center of the channel, if possible, so that air flows on both sides. This is more difficult to do and requires two glazing layers instead of one.

Construct the collectors carefully and insulate them well, particularly the upper areas that are likely to be hottest. Avoid insulations or glazings that will melt. If the collector's flow should be blocked for some reason on a sunny day, its temperature can reach over 300°F. Wood construction is usually satisfactory, but be sure to provide for wood shrinkage and for the expansion and contraction of materials as their temperatures fluctuates.

Systems with Heat Storage

Collectors that are large enough to supply more than 25 percent of the desired heat require heat storage. The storage in an air system is usually a large bin of rocks. It must be designed to maximize heat transfer from the air stream to the rocks without noticeably slowing the air flow. Rocks with small diameters (3 to 6 inches) have large amounts of surface area for absorbing heat, and yet allow passages for air flow. The rocks should be roughly the same size (that is, don't mix 1 inch with 4 inch) or most of the airways will be clogged. Storage should contain at least 200 pounds of rock (1½ cubic feet) per square foot of collector. As shown in the diagram, storage should be located

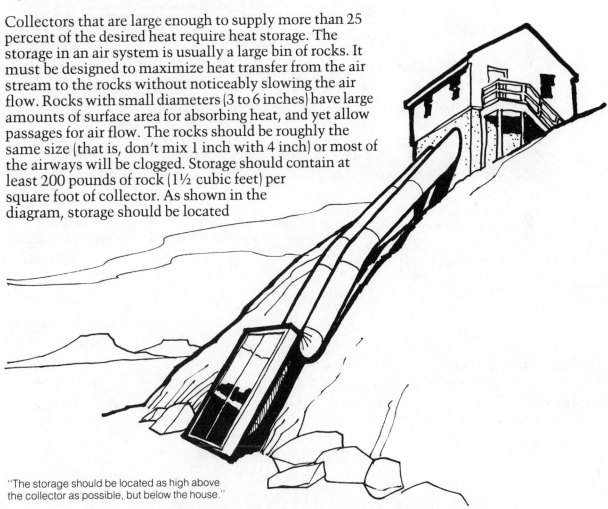

"The storage should be located as high above the collector as possible, but below the house."

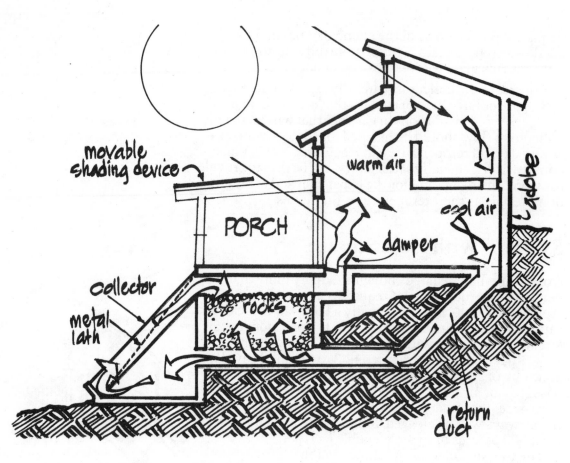

The cross-sectional area of the rock bed receiving air from the collector should range from one-half to three-quarters the surface area of the collector. The warm air from the collector should flow down through the rocks, and the supply air from storage to the house should flow in the reverse direction. Optimum rock size depends on rock bed depth. Steve Baer recommends gravel as small as 1 inch for rockbeds 2 feet deep and up to 6 inches for depths of 4 feet.* For best heat transfer in active systems, bed depths are normally at least 20 rock diameters. That is, if the rock is 4 inches in diameter, the bed should be at least 6½ feet deep in order to remove most of the heat from the air before it returns to the collector. This should be considered a maximum depth for convective loop rockbeds.

---

*See *Sunspots* by Steve Baer, Zomeworks Corporation, Albuquerque, NM.

above the collector but below the house. This permits solar heated air to rise into the house and cooler air to settle in the collector.

Air Flow

Designing a convective air loop system is a somewhat tricky and difficult task. If you aren't very respectful of the will of the air, the system won't work.

Steve Baer

As with active collectors, the slower the air flow, the hotter the absorber and the greater the heat loss through the glazing. This results in a lower collector efficiency. Good air flow keeps the absorber cool and transports the maximum possible amount of heat into the house. Flow channels should be as large as possible, and bends and turns in the ducts should be minimized to prevent restriction of air flow.

Conventional air heating collectors use fans and have air channels only ½ to 1 inch deep. Without fans, air channels in convective loop collectors range from 2 to 6 inches deep.

Convective flow of air is created by a difference in temperature between the two sides of the convective loops, for example, between the average temperature in the collector and the average temperature of the adjacent room. It is also affected by the height of the loop. The best air flow occurs when the collector is hottest, the room is coolest, and the height of the collector is as tall as possible.

The vertical distance to the top of the collector from the ground (this is not necessarily the collector length, since the collector is tilted) should be at least 6 feet to obtain the necessary effect. It should be tilted at a pitch of not less than a 45° angle to the ground, to allow for a good angle of reception to the sun and for the air to flow upward.

Reverse air convection

In an improperly designed system, reverse air convection can occur when the collector is cool. A cool collector can draw heat from the house or from storage. Up to 20 percent of the heat gained during a sunny day can be lost through this process by the following day.

There are three primary methods of automatically preventing reverse convection. One is to build the collector in a location below the heat storage and below the house. A second is to install backdraft dampers that automatically close when air flows in the wrong direction. One such damper is made of lightweight, thin plastic film. A lightweight "frisket" paper used in the photography industry has also been used successfully. Warm air flow gently pushes it open. Reverse cool air flow causes the plastic to fall back against the screened opening, stopping air flow. Ideally, both top and bottom vents should be equipped with such dampers.

This is discussed in more detail in two excellent magazine articles by W. Scott Morris (*Solar Age*, September 1978 and January 1979, Harrisville, N.H. 03450).

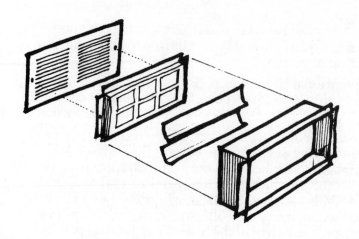

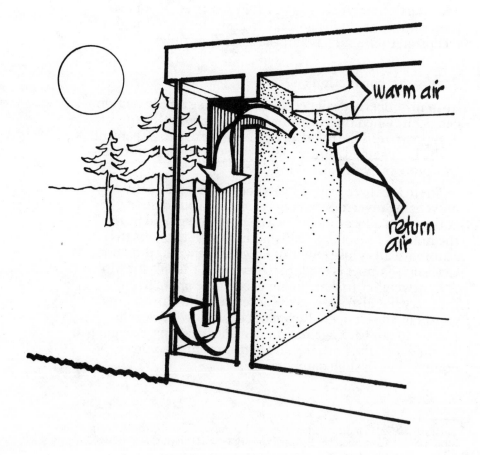

The third method of reducing reverse convection is to place the intake vent slightly lower than the outlet vent near the top of the wall. The back of the absorber is insulated and centered between the glazing and the wall. Inlet cool air from the ceiling drops into the channel behind the absorber. The solar heated air rises in the front channel, drawing cooler air in behind it. The warmed air enters the room at the top of the wall. When the sun is not shining, the air in both channels cools and settles to the bottom of the "U-tube." Only minor reverse convection occurs. Because of the longer air-travel distances involved, the U-tube collector will not be as efficient, aerodynamically, as the straight convective loop. It will also be more expensive to build.

Costs and Performance

Materials costs of solar chimney collectors can be as little as a few dollars per square foot. Materials are usually available locally. Contractor-built collectors can cost $7 to $15 per square foot. Operating costs are nonexistent, and maintenance costs should be very low.

Performance depends largely on delicate, natural convection currents in the system. Therefore, proper design, materials, and construction are important. In a well-built collector, air flow can be low to nonexistent at times of little or no sun, but will increase rapidly during sunny periods. Average collection efficiency is similar to that of low temperature, flat-plate collectors used in standard active system designs.

Collector Area

Only a small collector area is needed to heat a house in the spring and fall when the heating demand is low. Additional collector area provides heat over a fewer number of months, only during the middle of the heating season. Therefore, each additional square foot of collector supplies slightly less energy to the house than the previous square foot.

The useful amount of heat supplied from a solar collector ranges from 30,000 to 120,000 Btus per square foot per winter. The high numbers in this range are for undersized systems in cold, sunny climates. The low numbers are for oversized systems or for very cloudy climates. In cold climates of average sunshine such as Boston, Massachusetts, 80,000 Btus per square foot per heating season is typical, when the solar system is sized to contribute 50 percent of the heat. The output of the collectors drops to 50,000 Btus when sized to provide 65 to 70 percent of the heat. (For comparison, roughly 80,000 Btus are obtainable from a gallon of oil.)

Chapter 5

Solar Walls

Solar windows let sunlight directly into the house. The heat is usually stored in a heavy floor or in interior walls. Thermal storage walls, as solar walls are often called, are exactly what their name implies—walls built primarily to store heat. The most effective place to build them is directly inside the windows, so that the sunlight strikes the wall instead of directly heating the house. The directly sun-heated wall gets much hotter, and thereby stores more energy, than thermal mass placed elsewhere.

These "solar walls" conduct heat from their solar hot side to their interior cooler side, where the heat then radiates to the house. But this process takes a while. In a well-insulated house, a normal number of windows in the south wall will admit enough sun to heat the house during the day. Thermal storage walls will then pick up where the windows leave off and provide heat until morning.

South-facing windows with an area of less than 10 percent of the floor area of the house are probably not large enough to provide enough heat during the day. If this is the case, vents could be added at both the base and the top of a solar wall. The wall can then provide heat to the house during the day just as solar chimneys do. Although the vents need be only 10 to 12 square inches for each lineal foot of wall, they can add cost and complication. Therefore, it is best not to use them unless heat is needed during daylight hours. Thermal storage walls with vents are normally called Trombe Walls, after Dr. Felix Trombe who, in the early 1960s, built several homes with this design in the French Pyrenees.[1]

---

[1]Actually, the concept was originated and patented by E.L. Morse of Salem, Massachusetts, in the 1880s. His walls, complete with top and bottom dampers, used slate covered by glass.

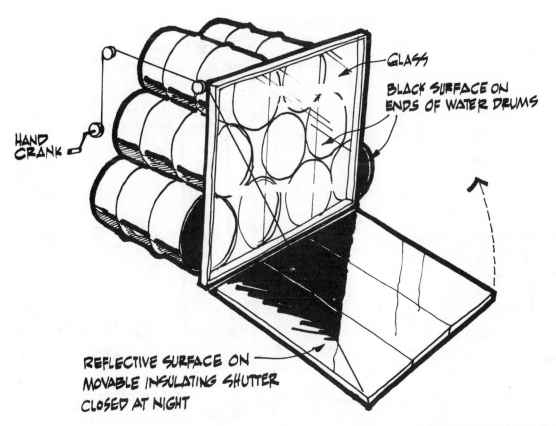

GLASS

BLACK SURFACE ON ENDS OF WATER DRUMS

HAND CRANK

REFLECTIVE SURFACE ON MOVABLE INSULATING SHUTTER CLOSED AT NIGHT

This water wall, called Drumwall™, was first developed by Steve Baer of Zomeworks Corporation. It consists of 55-gallon drums filled with water. Insulating panels hinged at the base of each wall cover the single layer of glass at night to reduce heat loss. With the panels open and lying flat on the ground, the aluminum surface reflects additional sunlight onto the drums. During the summer, the panels in the closed position shade the glass.

One type of thermal storage wall uses poured concrete, brick, adobe, stone, or solid (or filled) concrete blocks. Walls are usually one foot thick, but slightly thinner walls will do, and walls up to 18 inches thick will supply the most heat. Further thicknesses save no additional energy.

Containers of water are often used instead of concrete. They tend to be slightly more efficient than solid walls because they absorb the heat faster, due to convective currents of water inside the container as it is heated. This causes immediate mixing and quicker transfer of heat into the house than solid walls can provide. One-half cubic foot of water (about 4 gallons) per square foot of wall area is adequate, but unlike solid walls, the more water in the wall, the more energy it saves.

The main drawback of solar walls is their heat loss to the outside. Double glazing (glass or any of the plastics) is

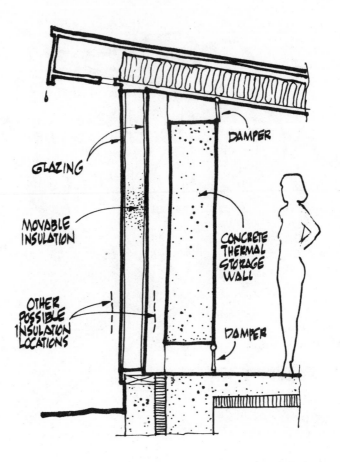

Movable insulation reduces heat loss from this concrete thermal storage wall. Hinged or sliding insulating shutters, reflective mylar roller shades, and other forms of movable insulation can be used. However, it is usually a tricky challenge to design the movable insulation systems so that its operation is simple and convenient. Insulating values of R-4 to R-6 will do.

The economic value of movable insulation in passive systems increases as the climate becomes more severe. However, most concrete storage wall systems do not use movable insulation because of its relative inconvenience and expense. Triple glazing is being used increasingly as a suitable alternative in cold climates.

adequate for cutting this down in most climates where winter is not too severe (less than 5000 degree days: Boston, New York, Kansas City, San Francisco). Triple glazing or movable insulation is required in colder climates.

Costs

Installation costs are affected by local construction practices, building codes, labor rates, and freight rates. Walls made of poured concrete and masonry block are less expensive in areas of the country where these materials are commonly used. The exterior glazing can be low in cost if an experienced subcontractor is available or if materials

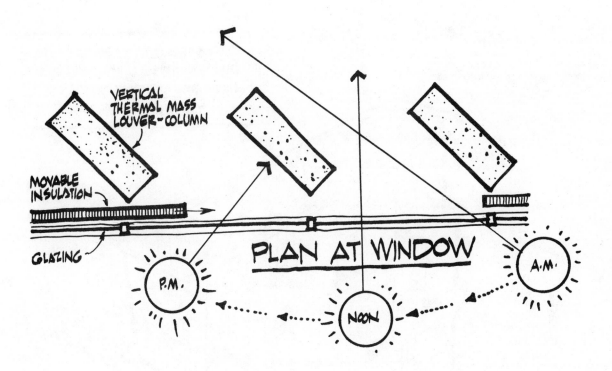

VERTICAL
THERMAL MASS
LOUVER-COLUMN

MOVABLE
INSULATION

GLAZING

PLAN AT WINDOW

P.M.

NOON

A.M.

An alternative to the solid concrete wall are Vertical Solar Louvers, a set of rectangular columns oriented in the southeast-northwest direction. They admit morning light into the building and store much of the heat from the afternoon sun. The inside of the glass is accessible for cleaning, and movable insulation can be easily installed between the glazing and the columns. The columns do, however, use precious living space. This variation of a solar wall was first used by Jim Bier.

Modules of cast fiberglass-reinforced polyester, from One Design, Inc., nest inside each other during transport. After the house is built, they are stacked atop each other and filled with water. Each module is about 8 feet long, 2 feet high, and 16 to 20 inches wide.

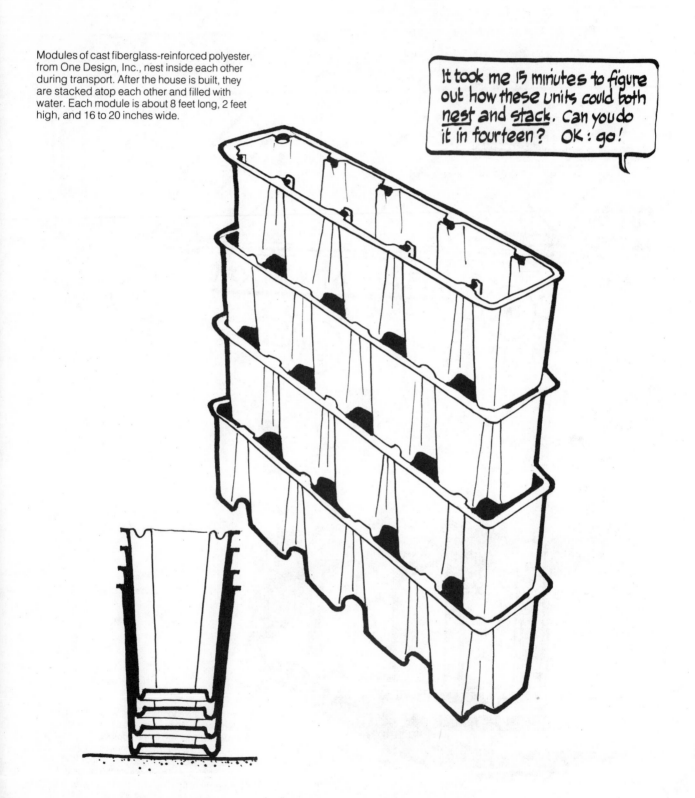

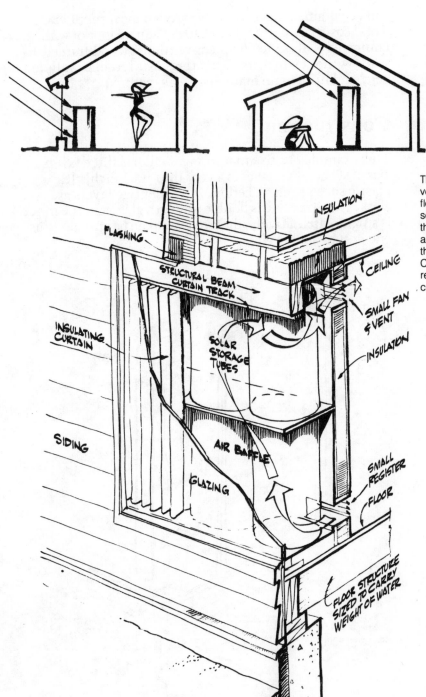

This thermal storage wall uses water-filled vertical tubes. To provide more control of heat flow into the house, a normal interior wall separates the water wall from the living space. A thermostatically-controlled fan circulates room air past the warm tubes. At night, an insulating thermal curtain is drawn across the glass. Corrugated, galvanized culverts and fiberglass-reinforced polyester tubes are the more commonly used cylinders.

can be obtained inexpensively through local suppliers. Total costs range from $5 to $20 per square foot of wall compared with $3 to $5 for conventional wood-framed walls without windows. Operating costs for solar walls are zero, and little or no maintenance is required.

Construction

This example of a thermal storage wall has three layers of glazing. The inner layer is a very thin (.001 inch) clear plastic sheet, such as Teflon. The other two layers are glass. The outer one is double-strength glass and the inner is single. Alternatively, the two layers can be purchased as one unit of double glass.

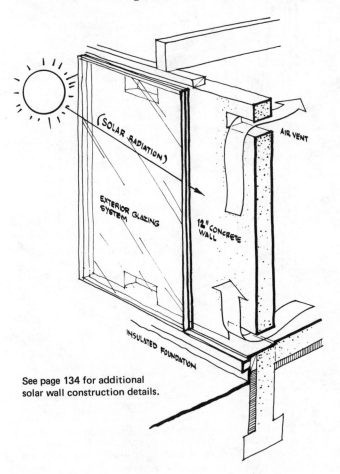

See page 134 for additional solar wall construction details.

Mount the entire glazing system one or two inches away from the wall. If the wall has vents, mount the glazing 3 to 4 inches away to allow for adequate air flow. Be sure to provide for the removal of cobwebs from the air space, and for cleaning and replacement of all glazing components. If you use aluminum, rather than wood, for framing and mounting the glass, place wood or other insulating material between the aluminum and the warm wall as a thermal separator. The glazing should extend above and below the face of the storage wall and be fully exposed to the sun. Glazing must be airtight and water resistant; it is the weather skin of the building.

The wall itself is of concrete, 12 inches thick, and of any height or width. It is either poured-in-place concrete, solid concrete block masonry, or concrete block filled with cement mortar. Use regular stone aggregate in the concrete (about 140 pounds per cubic foot). Lightweight aggregates should not be used, since they do not store as much heat. Unless the wall has to do structural work as well (such as supporting another wall or the roof), the concrete mix can be a relatively inexpensive one. When, as in many cases, the wall also supports the roof, reinforcing steel and structural anchors can be added without altering the wall's solar performance. In general, treat the juncture between the inner storage wall and the foundation floors, adjacent side walls, and roof as normal construction. However, make sure that house heat cannot easily escape through masonry or metal that is exposed to the weather. For example, foundations directly below glazed thermal storage walls should be doubly well-insulated from the ground.

If you install vents in the wall, use backdraft dampers to prevent reverse air circulation at night. (There are no commercial suppliers of these dampers, so see the previous chapter on solar chimneys for an example of one you can build.) Place the vents as close to the floor and ceiling as possible. Openings may be finished with decorative grills or registers. Such grills will keep inquisitive cats and tossed apple cores out of the airway, too!

Any interior finish must not prevent heat from radiating to the room. Just seal and paint the wall any color, or sandblast or brush the surface to expose the stone aggregate. A plastic skim coat or plaster can be used. Sheet

The Heating Effect

Heat loss through solar walls, even after days of cloudy weather, is not much worse than through conventional walls. The overall U-value of solar walls is 0.23. Due to their solar gains, they are net energy producers. (They are also *neat* heat producers.)

The temperatures of the heat supplied by solar walls is generally more moderate than that supplied by conventional heating systems. The lower temperatures tend to be more comfortable and less drying. Vented solar walls provide air at the ceiling level at 90°F to 100°F at air flows of approximately one cubic foot per minute per square foot of wall area.

The interior surface of a twelve-inch-thick wall reaches its warmest temperature in the early evening and then coasts from there by releasing stored heat. Temperatures range from 65° to 85° over the course of a day. The interior surface of a 24-inch wall peaks in temperature about 8 hours later. It doesn't get nearly as warm as thinner walls, but it provides heat more evenly for a longer period of time.

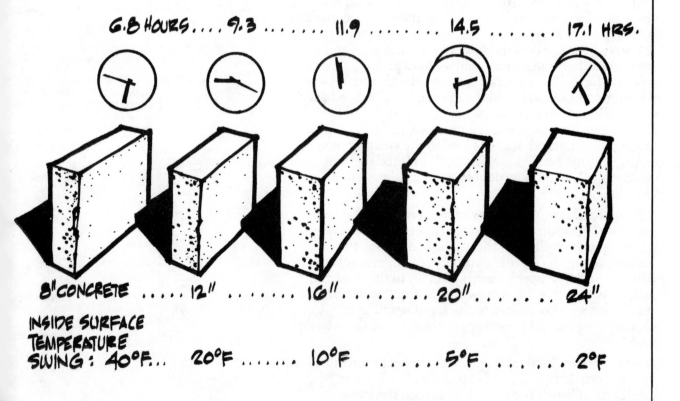

6.8 HOURS .... 9.3 .... 11.9 .... 14.5 .... 17.1 HRS.

8" CONCRETE ..... 12" .... 16" .... 20" .... 24"

INSIDE SURFACE TEMPERATURE SWING: 40°F... 20°F .... 10°F .... 5°F .... 2°F

materials, such as wood or hardwood paneling, should not be used. Use gypsum board only if excellent continuous contact between the board and the wall can be obtained—a difficult if not impossible task. Remember that many architects and interior designers regard natural concrete as an acceptable and attractive interior surface material.

Cleanse the solar (outer) surface of the wall with a masonry cleaner, prior to painting with virtually any dull-finish paint. Although dark brown and dark green have been used, flat black paint is preferred for maximum heat absorption.

Converting Your Existing Home

Solar walls are more difficult to add to an existing home than are solar windows, solar chimneys, and solar roofs. Uninsulated brick, stone, adobe, or block walls are candidates for conversion if they are unshaded during the winter and if they are oriented in a southernly direction (within 30° east or west of due south). Solid walls are more effective than walls that have air spaces, as is often found in brick walls comprised of two layers. If possible, the inner surface of the wall should be cleaned of conventional interior finish materials. Openings for vents are usually very difficult to make in walls that are candidates for conversion to solar walls. If, however, your windows are not large enough to supply all the heat you need during the day, and if heat is more important during the day than during the night (as for example, in an office building or a school), install vents of the sizes and in the proportions described earlier in this chapter for new walls. Paint the wall, and then cover it with the glazing system appropriate to your climate. For unvented walls, cover the wall first with an inexpensive sheet of plastic to bake the solvents out of the paint. When the plastic is coated with a thin film, remove the plastic and proceed with the installation of the permanent glazing system.

Summer Shading

A solar wall will supply a small amount of heat through the summer and have an effect on cooling bills. Shading the wall, with an overhang, an awning, or a tree, is the most effective method to reduce its exposure to direct sunlight. A cloth or canvas draped over the wall is also effective.

Some people choose to place vents in the framing at the top of the wall. The vents are open during the summer, permitting the heat from the wall to escape to the outside. Their primary shortcoming is that they are prone to leakage during the winter. This air leak can have a significant effect on the performance of the wall during the winter.

The earlier illustration of water drums has a movable insulating shutter that lies flat in front of the wall. In addition to its functions as a solar reflector and heat insulator during the winter, it can act as a shading device during the summer when it is in the vertical closed position.

Chapter 6

Solar Roofs

Solar roofs, often called thermal storage roofs, are similar to storage walls. Waterbed-like bags of water, exposed to sunlight, collect, store and distribute heat. This heat passes freely down through the supporting ceiling to the house, gently warming it. In the summer, heat rises through the ceiling into the water, cooling the house. Then at night, the water is cooled by the radiation of its heat to the sky. Movable insulation covers the ponds at night in winter, to trap heat inside, and during the day in summer, to shade the ponds while the sun is shining. See page 71 for a diagram of the system.

Generally, solar roof ponds are 8 to 12 inches deep. Roof ponds are always flat, but in northern buildings the glazing is often sloped to the south to capture the sun's low rays as well as to shed snow. Under the sloped glass, the walls are well-insulated and faced with materials that reflect the sunlight into the ponds.

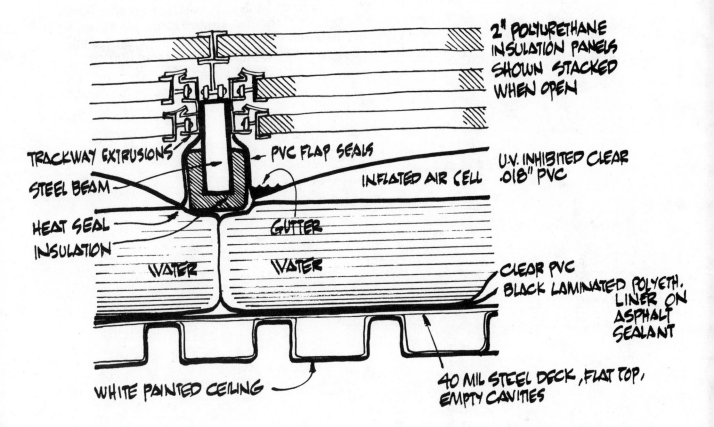

2" POLYURETHANE INSULATION PANELS SHOWN STACKED WHEN OPEN

TRACKWAY EXTRUSIONS

PVC FLAP SEALS

U.V. INHIBITED CLEAR .018" PVC

INFLATED AIR CELL

STEEL BEAM

HEAT SEAL INSULATION

GUTTER

WATER

WATER

CLEAR PVC BLACK LAMINATED POLYETH. LINER ON ASPHALT SEALANT

WHITE PAINTED CEILING

40 MIL STEEL DECK, FLAT TOP, EMPTY CAVITIES

This cross section is of the solar roof system used in a house designed by Harold Hay in the mild climate of Atascadero, California. The entire 1100-square-foot ceiling is covered with 8 inches of water sealed in clear UV-inhibited, 20 mil, polyvinylchloride water bags. Underneath these 53,600 pounds of water is a layer of black polyethylene to help absorb solar radiation near the bottom of the bags. Additionally, an inflated clear plastic sheet above the water bags enhances the "greenhouse (or heat trapping) effect" during the heating season. This air cell is deflated in the summer months to permit radiational cooling. A 40 mil steel deck roof supports the water bags and provides good heat transfer to and from the living space. Above the roof ponds, a system of movable insulating panels is mounted on horizontal steel tracks. The insulation is 2 inches of rigid polyurethane faced with aluminum foil. The panels are moved by a 1/6 horsepower motor operating about 10 minutes per day.

Solar roof ponds maintain very stable indoor temperatures. During the winter and summer, temperatures typically fluctuate between 66° and 73° while the outdoor average daily temperature fluctuates between 47° and 82° throughout the entire year. The Atascadero house is 100 percent solar heated and cooled, and it has no other source of heating or cooling. Occupants have found the heating and cooling system provides "superior" comfort compared with conventional systems.

Not many solar roofs have been built, and there is limited information on the design, cost, performnce, and construction details of thermal storage roofs. However, they offer tremendous potential for reducing heating and cooling bills. Page 126 has photos of solar roof houses.

Chapter 7

Solar Rooms

Without ventilation or thermal mass, the temperatures of spaces having large areas of south-facing windows will fluctuate widely. Temperatures of conventional nonsolar greenhouses, for example, can rise to over 100° on sunny winter days and then drop to below freezing at night. If a sunheated room is permitted to have wider-than-normal temperature fluctuation, then the costs of thermal mass (to store heat) and movable insulation (to reduce heat loss) are avoided. The excess warmth from such a "solar room" can heat the house immediately, or if mass is added, heat can be stored for later use after the sun sets. Almost always, the solar room is warmer than the outdoor temperature, thus reducing heat loss from the building where the room is attached. Examples of solar rooms include greenhouses, solariums, and sun porches.

Greenhouses are the most common solar rooms. Conventional greenhouses, however, are not designed to take maximum advantage of the sun's energy. The problem is that most are built with a single layer of glass, and so they lose more heat at night than they gain from the sun during the day. Consequently, they need expensive auxillary heat to keep the plants warm.

A solar greenhouse is designed both to maximize solar gain and to minimize heat loss. Usually, only the south-facing walls and roof of the solar greenhouses are glazed, while the east and west walls are well-insulated. (Southeast and southwest portions, if any, are also glazed, partly because plants need that low-angle early sunlight.) If at least two layers of glass or plastic are used instead of one, this type of greenhouse will remain above freezing most of the winter in all but the coldest climates of this

Which Direction?

Solar rooms that face east or west do not work as well for heating as those that face south. The former supply less heat during the winter and may provide much too much in summer. However, an east-facing greenhouse can give morning light, which plants like; it can be a buffer zone to reduce heat loss from the house throughout the rest of the day. If an east-facing solar room seems to be a good solution to either site or building problems, locate spaces such as kitchens on the east side of the house next to or behind the solar room to take advantage of the morning light and heat. Then the living rooms and bedrooms, which can usually remain cool during the day, will become warm in the afternoon from the heat gained from the west.

country. However, for maximum heat savings while growing plants year round, three and even four layers of glass and plastic should be used where winters have more than 5000 degree days. Keep in mind that each additional layer of glazing blocks additional sunlight. Therefore, for the highest possible light transmission, the third and fourth layers must be a very clear film, such as Teflon™ or Tedlar™. Each layer must be sealed tightly to prevent structural damage from possible moisture condensation between glazings.

For maximum sunshine, and for minimum heat loss at night, movable insulation is used in combination with double glazing. This can be tough to do, however. Some of the tricky design and construction problems include storing the insulation out of the way during the day,

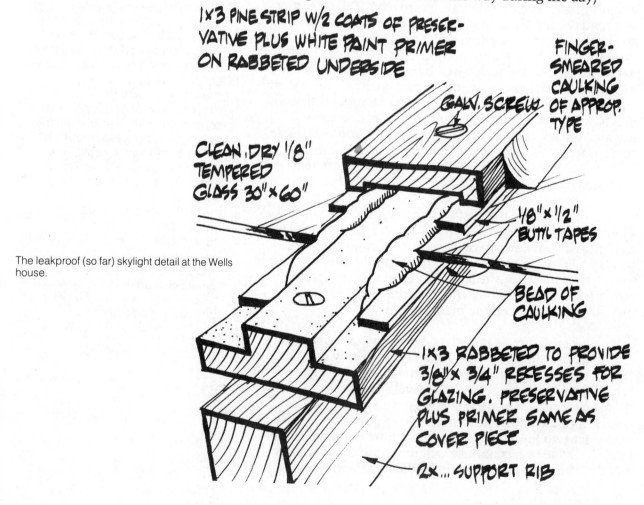

The leakproof (so far) skylight detail at the Wells house.

interfering with plants while moving the insulation, and obtaining tight seals against the glazing when the insulation is closed. Additional considerations include the need for insulation to resist mold, other plant and insect life, and moisture damage.

Glazing for solar rooms should be vertical or sloped no more than 30° from vertical (at least 60° from horizontal). Before you build, however, talk to everyone you can find who has ever used glass in a sloping position, and ask about leaks. If you can find someone who can convince you of a leakproof system, do not let any details escape your attention. Also, read the fine print in the sealants literature. Some silicones are attacked by mildew, many won't stick to wood, and all must be applied only to super clean surfaces.

Heat Storage

As with other passive systems, thermal mass enhances the performance of a solar room. Thermal storage mass moderates temperature swings, provides more stable growing temperatures for plants, and increases overall heating efficiency. The heat-storing capability of the planting beds can be supplemented with 55-gallon drums, plastic jugs, or other containers of water. Two to four gallons of water per square foot of glazing is probably adequate for most solar rooms.

Many of the most successful solar rooms are separated from the house by a heavy wall that stores the heat. The wall, built of concrete, stone, brick, or adobe, conducts heat (slowly) into the house. At the same time, the wall keeps the solar room cooler during the day and warmer at night. Use the design and construction information for solar walls, but eliminate the glazing.

Earth, concrete, or the floors store considerable heat. So do foundation walls if insulated on the outside. Be sure to use insulation with an R-value of at least 12 (3 inches of styrofoam™). Insulate at least 3 to 4 feet deep and more in deep-frost country. This gives better protection than insulating 2 feet or so horizontally under the floor.

Selecting the Right Glazing Material

Appendix 6 provides assistance in choosing the right materials for glazing solar rooms. Column three of Table 1 in Appendix 6 states the solar transmittance of each material listed. This figure represents the fraction of sunlight that actually passes through the glazing. To find the transmittance of several layers, multiply together (don't add) the transmittance of all of the materials. For example, for a glazing system of two layers of window glass (transmittance of 0.91—see line three) and two layers of Tedlar™ (transmittance of 0.95—see line seven), the total transmittance is

$$0.91 \times 0.91 \times 0.95 \times 0.95 = 0.75.$$

Do not use glazing systems that have a transmittance of less than 0.70. Lower light levels will jeopardize plant growth.

When solar rooms larger than 200 square feet reach 90°, a fan can be used to circulate the collected heat. Because plants benefit from having warm soil, hot air can be blown horizontally through a 2-foot-deep bed of stones below the greenhouse floor or under raised planting beds. Stone beds can also be built beneath the floor of the house and should not be insulated from it. Then the heat will rise naturally through the stone beds and into the planting bed soil or into the house.

Two cubic feet of ordinary washed stone per square foot of glazing is sufficient. Use a fan capable of moving about 10 cubic feet of air per minute for each square foot of glazing. Potato-sized stones, larger than the usual ¾ inch

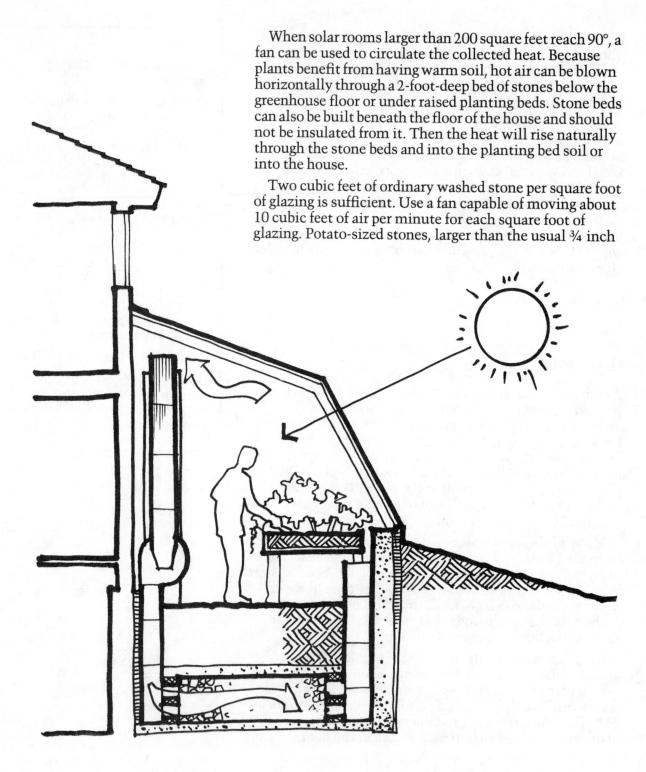

to 1½ inch size, will allow freer air movement. Consult with a local mechanical engineer or heating contractor for the best fan and ducting design. (Keep it simple!)

Costs

Solar rooms can be relatively simple to build, yet they can be very expensive if they are of the same quality and durability as the rest of your house. For example, with a few hundred dollars worth of materials, you can build a simple, wood-framed addition to your house to support thin-film plastics. The resulting enclosure will provide considerable heat, especially if it is not used for growing plants. On the other hand, good craftsmanship and quality materials can result in costs of several thousand dollars. In general, solar rooms are most economical when you can use them for more than providing heat and when they are built to a quality that will both enhance the value of your house and appreciate in value as your house does. Solar rooms are often exempt from local property taxes. Check with your local officials.

Large Solar Rooms

Most of the information in this chapter is applicable for relatively small solar rooms of 100 to 200 square feet. Unless your house is super insulated or in a mild climate, a solar room of this size will provide less than 25 percent of your heat. For big leaky houses, small solar rooms will provide as little as 5 or 10 percent of the heat.

Another way of approaching the use of solar rooms to heat your house is to think of them as rather large spaces that are incorporated into, rather than attached onto, your house. There are a number of advantages with this approach:

1. Both the solar room and house will lose less heat.
2. Heat will move easily from the solar room to your house.
3. Natural light can be made to penetrate deep into your house.

Growing Plants; some things to remember

An important function of some solar rooms is the growing of food—and flowers. Warm soil and sufficient light are critical for successful plant growth. Remember that the multiple layers of glass or plastic you may need to use will reduce light levels, a crucial issue in climates with below-average sunshine. Circulation of warm air through gravel beds under the soil can raise planting bed temperatures, increasing the growth rate of most plants.

Cold weather plants, such as cabbage, can tolerate cold temperatures, sometimes even mildly-freezing ones. Few house plants can be permitted to freeze, but many can endure rather cool temperatures. On the other hand, some plants, such as orchids, require stable, high temperatures. When warm, stable temperatures are required, the solar room must retain most of its solar heat; little heat should be allowed to move into the house. Three or four layers of glass or plastic (or movable insulation) and plenty of thermal mass are required to trap and contain the heat in cold climates.

Evaporation of water from planting beds and transpiration by the plants causes humidity. Each gallon of water thus vaporized used roughly 8000 Btus, nearly the same amount of energy supplied by 5 square feet of glass on a sunny day. Also, water vaporization reduces peak temperatures. It may be undesirable to circulate moisture-laden air into the house, unless the house is very dry.

Greenhouse environments are rather complex ecological systems. Unexpected and sometimes undesirable plant and animal growth may proliferate. Indications are that the greater mix of plants and animals, the more likely a natural balance will eventually be reached. To obtain this balance, some owners leave the door of their solar room open to the outside during the warm and mild weather.

New Alchemy Institute, among others, has pioneered work in natural pest control and companion planting as a step toward successful greenhouse management. They have also investigated fish-raising in large "aquariums," which also serve as thermal mass. Human, animal, and/or plant wastes are integrated into the total ecology of many advanced greenhouses, which are sometimes referred to as bioshelters. A more thorough understanding of the many natural cycles that are possible in greenhouses will offer rewards.

*New Alchemy Institute, E. Falmouth, MA 02536.

4. The solar room can be easily heated by the house if necessary and so is unlikely to freeze.

5. The solar room can be readily used as an expanded living space.

6. You can build your house compactly and the solar room will provide a feeling of large exterior wall and window area.

Attic Solar Rooms

Attics are often great places for solar rooms, particularly if their only purpose is to heat your home. Frame the roof in a conventional manner. Glaze the south slope with one sheet of glass or plastic. Insulate the end walls, the north roof, and the floor. Paint all of the surfaces black. When your house needs heat and the solar room is hot, a fan can circulate solar heated air from the attic to the house.

Be sure to insulate the sun-trap from the rest of the house. Place back-draft dampers on the air ducts to prevent house air from rising up into the attic at night when the attic is cold and the house is warm. This solar room design gets very cold on winter nights but heats up quickly when the sun shines.

Because it has no thermal mass to store the heat that the house doesn't need, it is unable to reduce fuel bills by more than 20 to 25 percent. In order to reduce fuel bills further, the design must be altered in a number of ways. First, the glazed portion of the roof must exceed 15 to 20 percent of the floor area of the house.

Second, thermal mass must be added to the attic. This is frequently done by placing containers of water along the north wall of the attic.

Third, movable insulation must cover the glazing at night to significantly reduce heat loss from the attic so as to trap and store the sun's heat. In climates of 3,000 degree days or less, double or triple glazing is an alternative to movable insulation. Kalwell Corporation (see p. 161 of the appendix) has developed detailed construction and design information.

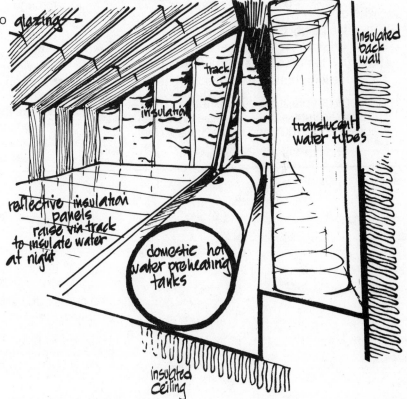

Ventilation

Even the best designed solar rooms will require ventilation at times of too much heat or humidity or too little carbon dioxide. Ventilation should be able to replace all of the room's air up to six times each hour.

Natural ventilation is preferred to energy-consuming mechanical ventilation. The greatest amount of ventilation occurs when the exhaust vents are positioned as high as possible and the intake vents as low as possible. Air flow rates and, in turn, necessary vent sizes, can be estimated. The velocity of the air, in feet per minute is approximately

$$V = 486 \sqrt{\frac{h(T_0 - T_i)}{T_i + 460°}},$$

where

h = the height between the intake vent and exhaust vent,

T_0 = the temperature at the outlet vent, and

T_i = the temperature at the intake vent.

For example, if the outdoor temperature at the intake vent is 85°, the temperature at the outlet vent is 100°, and the height is 10 feet,

then the velocity is

$$V = 486 \sqrt{\frac{10(100 - 85)}{85 + 460}}$$

= 255 cubic feet per minute (255 cf/min).

For solar rooms that taper at the top (as in lean-tos), smaller values should be used. Air can carry 0.018 Btu per cubic foot for each degree it increases in temperature (0.018 Btu/cf−°F). The amount of heat exhausted through one square foot of vent each hour is, therefore

(255 cf/min) × (100° − 85°)
 × (0.018 Btu/cf-°F) × 60 min) = 4131 Btus.

A representative value for solar heat gain through glass is 200 Btus per square foot each hour. Therefore, each square foot of vent can accommodate 20 square feet of glass.

Solar rooms must sometimes be ventilated in midwinter, when heat from the bright sunlight builds up too quickly to be dissipated through the house or absorbed by thermal mass. Just be sure that such vents are sealed tightly on winter nights, and on cold sunless days.

7. The costs can be less than for solar rooms that are simply added onto conventional house designs.

8. The excess humidity of the solar room can be somewhat reduced by, and profitably used by, the exclusively dry winter house.

Perhaps the most notable example of this approach to solar rooms is the Balcomb residence in Santa Fe, New

How to Get the Heat from the Solar Room into your House

A. Windows

Windows in the walls between the solar room and the house let light pass right into the house (just as in direct gain systems), especially during winter months when the sun is low in the sky. The roof of the solar room can shade the windows during the summer, helping to keep the house cool. Since the house windows are protected from the weather, you can keep their construction simple and inexpensive.

B. Natural Air Movement

When your solar room is warm, just open the windows and doors and let the heat flow into the house. The higher the windows or other openings, the more heat will flow inside. And you can use curtains to control the flow of heat. However, don't forget about odors, insects, and humidity. Screens over the openings are usually a must. A fan on a simple thermostat can regulate the amount of air flow into the house. The small extra expense will ensure that your "solar system" works when you're not home.

C. Conduction

Conduction through an unglazed thermal storage wall is one of the best ways of transferring heat into your house. The wall should not be insulated. In the summer, the wall will protect the house from the solar room's heat. If the wall is wood-framed, it should be insulated. Be sure to protect the wood from the moisture of the solar room.

D. Gravel Beds

Use fans to blow warm air from the solar room through gravel beds under uninsulated floors of your house. Heat will radiate up through the floor into the room that is to be heated. The fans can be kept off during mild weather. Do not use fans to circulate the air from the gravel beds directly into the house as there could be dampness and musty-smell problems. Radiant heat through the floor is much more effective and comfortable.

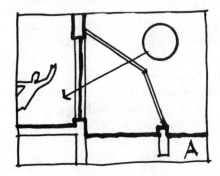

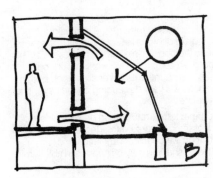

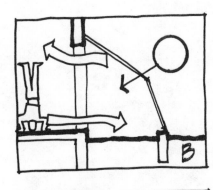

Mexico (see p. 124). The long edge of the two-story triangular-shaped sunspace faces south. The house wraps around the northeast and northwest faces of the triangle. The annual electric heating bill is less than $100.

Another example of this approach is the "Solar Envelope" house developed by Lee Porter Butler (see p. 125). The entire south side of the house is a two-story solar room. Warmed, greenhouse air rises through a roof plenum down a plenum in the north wall, through the crawl space or cellar, and back through the greenhouse. The roof and wall plenums extend the full east-to-west length of the house and are, in effect, a cavity or "envelope" buffering the house from the extremes of the outdoor weather. Both faces of the plenum are well-insulated.

During the summer, the excess heat from the solar room is vented at the peak of the house to the outside, helping to ventilate the house and, in some climates, pulling the outside air through tubes, buried in the earth, that dehumidify and cool the air. Many houses of this design have heating and cooling bills of less than $100 per year.

An All-Purpose Solar Room Design
It is difficult to sort through the confusing multitude of designs for solar rooms and to choose the one that makes the most sense. This all-purpose solar room will work throughout most of the country. Its net energy contribution to a house will vary depending on the severity of the climate.

Summer temperatures can be kept close to outdoor temperatures with adequate ventilation. Mechanical ventilation and/or additional shading may be needed in hot, humid climates.

Winter temperatures in the solar room are likely to be as follows:

Up to 8000 degree days and more than 70% possible sunshine: 45°–85°

Up to 8000 degree days and less than 70% possible sunshine: 35°–85° with occasional need for backup heat

More than 8000 degree days and more than 70% possible sunshine: 35°–85° with occasional need for backup heat

More than 8000 degree days and less than 70% possible sunshine: up to 85° with frequent need for backup heat.

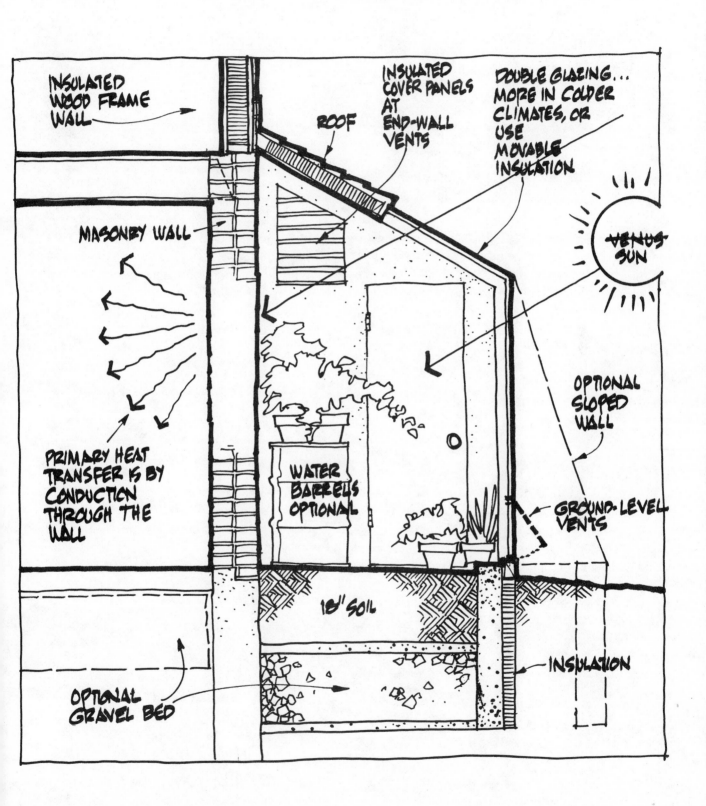

INSULATED WOOD FRAME WALL

ROOF

INSULATED COVER PANELS AT END-WALL VENTS

DOUBLE GLAZING... MORE IN COLDER CLIMATES, OR USE MOVABLE INSULATION

VENUS SUN

MASONRY WALL

OPTIONAL SLOPED WALL

PRIMARY HEAT TRANSFER IS BY CONDUCTION THROUGH THE WALL

WATER BARRELS OPTIONAL

GROUND-LEVEL VENTS

18" SOIL

INSULATION

OPTIONAL GRAVEL BED

PART 3

Passive Solar Cooling

The air-conditioned suit was a great economical idea of the age—cool in summer and warm in winter. (Woodcut by Bellows, 1877. The Bettmann Archive.)

Chapter 8

How the Sun can Cool

"What do you mean *solar* cooling?" That's a good question. In fact, of the several passive solar cooling techniques discussed here, only one of them can be labeled strictly "solar," and even at that only if you stretch the definition. But all the techniques use passive, or natural, cooling that require no pumps or fans for their operation. Many of them are based on plain old common sense. Fortunately, in virtually all climates, houses can be designed to provide ideal comfort without mechanical air conditioning. This is not to say it is easy to do everywhere. In really hot and humid climates where electricity is still inexpensive, it just might not be worth the extra effort; not yet, but it may be someday. And it's at least nice to know such things are possible.

By far the first and most important step in cooling is to keep sunlight from falling on your house, so first we'll discuss "solar control." We'll also discuss natural cooling by ventilation, evaporation, sky radiation, and nesting of buildings into the earth.

Solar Control

Usually the easiest, most inexpensive and effective way to "solar" cool your house is to shade it—keep the sun from hitting your windows, walls, and roof. In fact, where summer temperatures average less than 80°, shading may be all you need to stay cool.

Most of the things you do to reduce winter heat loss also reduce summer heat gain. For example, heavily insulated walls keep out summer heat, so shading them is not as important as if they were poorly insulated. Facing windows south to catch the winter sun and minimizing east and west windows to reduce heat loss are also important steps in solar control. South windows admit less sun in the summer than they do in the winter, while east and west windows can turn houses into ovens. A light-colored roof may be 60 to 80 degrees cooler than a dark roof because it reflects light.

Reducing the Need for New Power Plants (No More Nukes)

Natural cooling can significantly reduce peak cooling power loads. With natural cooling, the size and use of backup conventional electrical air conditioners is reduced. This means the demand for power is reduced and so is the need for new power plants. For example, approximately 50 square feet of west window unshaded from the sun needs a 1-ton air conditioner and approximately 2 kilowatts of electrical generating capacity costing $2500 or more to build. Shading the windows or facing them in other directions can reduce those peak power demands created during the summer months.

Thermal mass can absorb heat during the day, delaying the need for cooling until after peak demand hours. At that later time, the need will not be as great and the air conditioner can be smaller and cheaper to run.

In really hot climates, uninsulated walls and roofs should be shaded. But although this is important, shading windows is far more important. Overhangs and awnings work well. Unfortunately, fixed overhangs provide shading that coincides with the seasons of the *sun* rather than with the *climate*. The middle of the sun's summer is June 21—the longest day, when the sun is highest in the sky. But the hottest weather occurs in August, when the sun is lower in the sky. A fixed overhang designed for optimal shading on August 10, 50 days after June 21, also causes the same shading on May 3, 50 days before June 21. The overhang designed for optimal shading on September 21, when the weather is still somewhat warm and solar

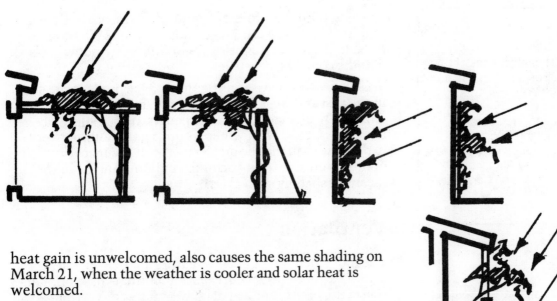

heat gain is unwelcomed, also causes the same shading on March 21, when the weather is cooler and solar heat is welcomed.

Shading from deciduous vegetation more closely follows the climatic seasons, and therefore, the energy needs of houses. On March 21, for example, there are no leaves on most plants in north temperate zones, and the bare branches readily let sunlight through. On September 21, however, those same plants are still in full leaf, providing needed shading.

Operable shades are even more versatile and adaptable to human comfort. The most effective shades are those mounted on the outside of a building. However, most exterior operable shades do not last very long. Nesting

The most significant sources of technical detail on shading can be found in the *ASHRAE Handbook of Fundamentals* by the American Society of Heating, Refrigerating, and Air Conditioning Engineers, New York, and in *Solar Control and Shading Devices* by Aladar and Victor Olgyay, Princeton University Press.

animals, climbing children, wind, and weather will see to that. Inside shades last longer, but few are as effective as outside shades. Once the sunlight hits the window glass, half the cooling battle is lost.

East and west glass is difficult to shade because the sun in the east and west is low in the sky in both summer and winter. Overhangs prevent the penetration of sunlight through east and west windows during the summer very little more than they do during the winter. Vertical louvers or other vertical extension of the building are the best means of shading such glass.

Ventilation

The movement of room temperature air, or even slightly warmer air across our skin causes a cooling sensation. This is because of the removal of body heat by convection currents and because of the evaporation of perspiration.

The most common way to cool a house with moving air without using mechnical power is to open windows and doors. Do not forget this simple concept—natural ventilation—and do not underestimate its cooling effect. Low, open windows that let air in result in air flow through

Natural ventilation can be affected by land planning. Natural breezes should not be blocked by trees, bushes, or other buildings. Shade trees should be selected so that branches and leaves are as high above the house as possible to allow a breeze to enter below them. The shape of your house, proper clustering of buildings, and other landscaping features such as bushes and fences can funnel and multiply natural breezes.

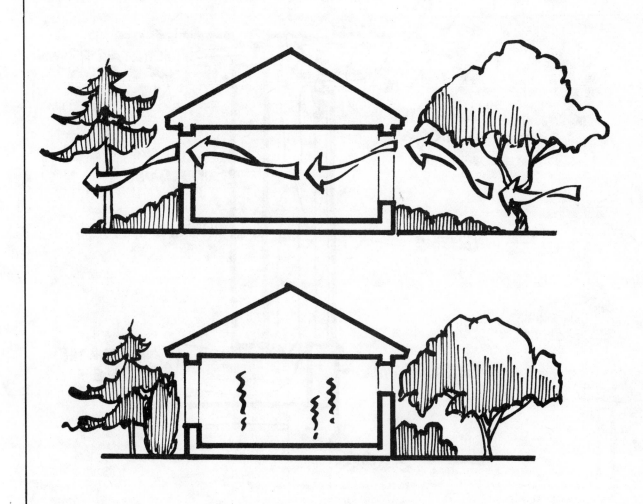

Here, a solar collector exhausts its hot air to the outdoors by natural convection and pulls house air through itself, providing ventilation. The solar collector is very similar to the solar chimney, convective loop collectors discussed in Chapter 4. Many variations of this "solar chimney" have been used widely in the past and are being developed again today. In some *active* solar systems using heated air (rather than a liquid), the collectors vent hot air to the outside during sunny summer weather, pulling house air through themselves, with or without the use of blowers.

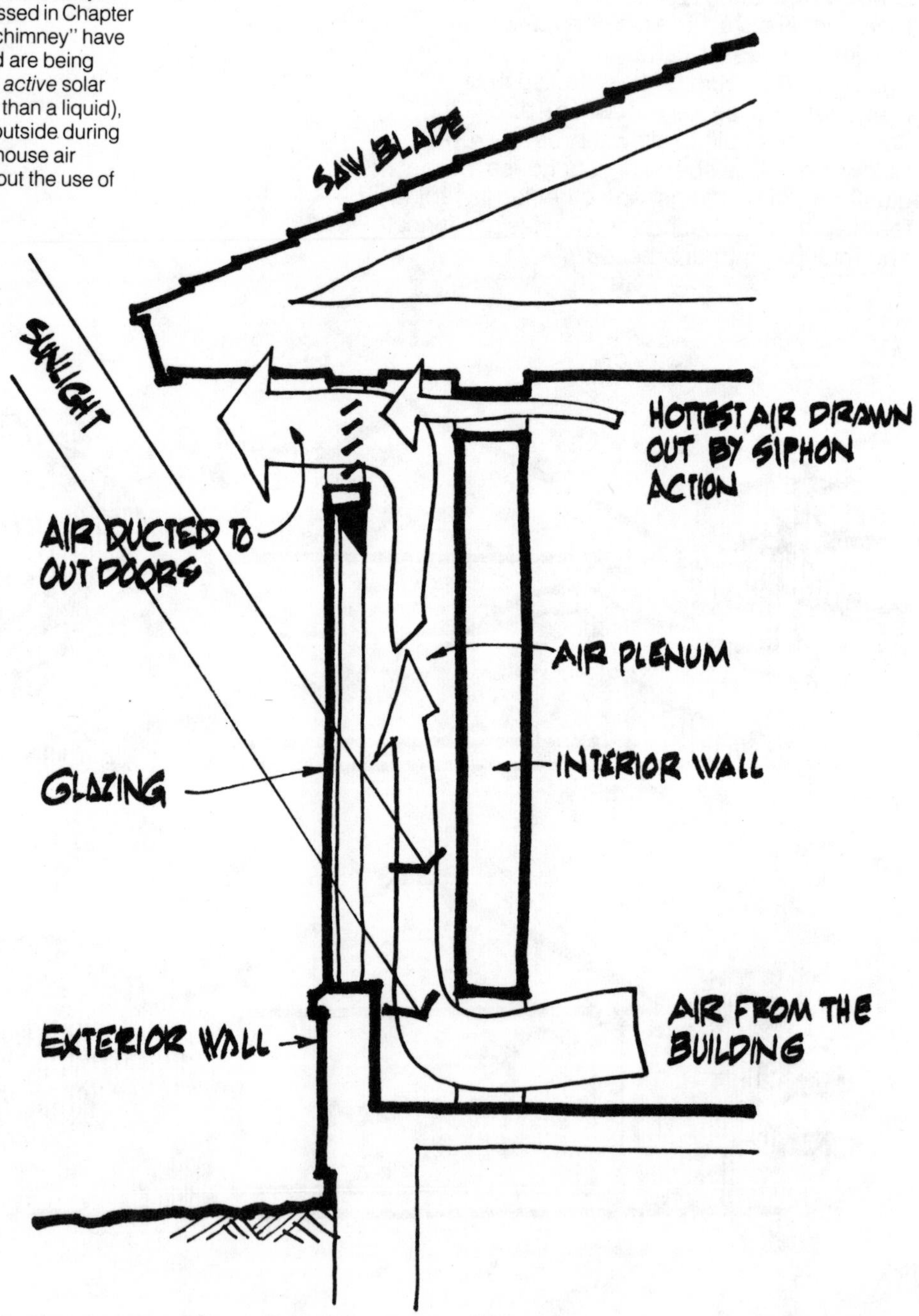

SAW BLADE

SUNLIGHT

HOTTEST AIR DRAWN OUT BY SIPHON ACTION

AIR DUCTED TO OUT DOORS

AIR PLENUM

INTERIOR WALL

GLAZING

EXTERIOR WALL →

AIR FROM THE BUILDING

the lower part of the room where people are, rather than near the ceiling. Houses that are narrow and face the wind, or that are T- or H-shaped, trap breezes and enhance cross ventilation through the house. When all else fails, open or screened porches, located at or near the corners of houses, can capture soft, elusive breezes as they glide around the house.

The "stack" or "chimney" effect can be used to induce ventilation even where there is no breeze. Warm air rises to the top of a tall space, where openings naturally exhaust the warm air. Openings at floor level let outdoor air in. Natural ventilation can be further induced by the use of cupolas, attic vents, belvederes, wind vanes, and wind scoops.

If your summer nights are a lot cooler than days, build your house of heavy materials. Cool the house at night by natural ventilation and the thermal mass will keep it cool during the day.

Determining Ventilation Air Flow

In stack-effect ventilation, air flow is maximized by the height of the stack and the temperature of air in the stack. Air flow is proportional to the inlet area and to the square root of the height times the average temperature difference, as follows:

$$Q = 540A \sqrt{h(T_1 - T_2)},$$

where

Q = the rate of air flow, in cubic feet per hour;

A = the area of the inlets, in square feet;

h = the height between inlets and outlets, in feet;

T_1 = the average temperature of the air in the "chimney," and

T_2 = the average temperature of the return air (normally just the outside temperature),

It is better to add heat (presumably using a passive air-heating collector) at the bottom of the chimney or stack than at the top. In this way the entire column of air in the chimney is hot, creating the desired bouyancy to cause the air to flow.

If outlet sizes are appreciably different from inlet sizes, the above expression must be adjusted according to the following ratios:

Area of Outlets / Area of Inlets	Value to be substituted for 540 in above expression
5	745
4	740
3	720
2	680
1	540
¾	455
½	340
¼	185

This information is from *Design with Climate* by Victor Olgyay, Princeton University Press.

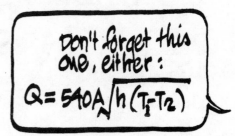
Don't forget this one, either:
$$Q = 540A \sqrt{h(T_1 - T_2)}$$

Evaporation

Our skin feels cooler if it's wet when the wind hits it than when it's dry due to the evaporation of moisture. The same process can cool air effectively, especially in dry climates. There are many ways to evaporate water. The most effective are by having large water surfaces, and by agitating, spraying, or moving water in contact with the air for the greatest surface contact. For example, large shallow ponds provide this large surface contact. Moving streams and sprays from water fountains increase turbulence and, thus, surface contact.

Evaporative cooling can also keep roofs cool. Roof sprays and ponds have been used successfully in hot climates such as Arizona and Florida. Wind-flows across the roof ponds should be enhanced, if possible, to encourage evaporation. This can be done mechanically, but careful design of roof shapes can help speed the natural flow of air across the ponds.

The transpiration of indoor plants has a cooling effect. So do interior pools and fountains. Fans can add

enormously to the evaporative effect and have been used successfully in "swamp" coolers and other evaporative coolers.

If you live in a humid climate, however, do not expect evaporative cooling to help you very much. In fact, it is likely to make your humidity problems worse.

Radiational Cooling

Thermal energy is constantly being exchanged between objects that can "see" each other. More energy radiates from the warmer object to the cooler object. The sun radiates heat to the earth, and the earth radiates considerable heat to clear night skies, which, even during hot weather, are quite cool. The northern sky is often cool during the day.

Most sky radiation occurs at night. The amount varies greatly from one part of the sky to another, from 100 percent possible directly overhead to virtually none at the horizon. The most effective radiant cooling surface is horizontal, facing straight up. Obstructions such as trees and walls reduce night sky cooling. A vertical surface with no obstructions yields less than half of the radiant cooling of an unobstructed horizontal surface.

The classic sky radiation cooling concept is Harold Hay's "skytherm" house in Atascadero, California. (Read Chapter 6 on Solar Roofs to learn more about this cooling effect.) To date, the use of sky radiation for cooling houses has been limited to climates with clear skies. Clouds and air with high moisture content significantly reduce the radiation of heat to the sky.

One of the best sources of engineering data and analyses of radiational cooling is "Atmospheric Radiation Near the Surface of the Ground: A summary for Engineers" by Raymond W. Bliss, *Solar Energy*, July/September 1961.

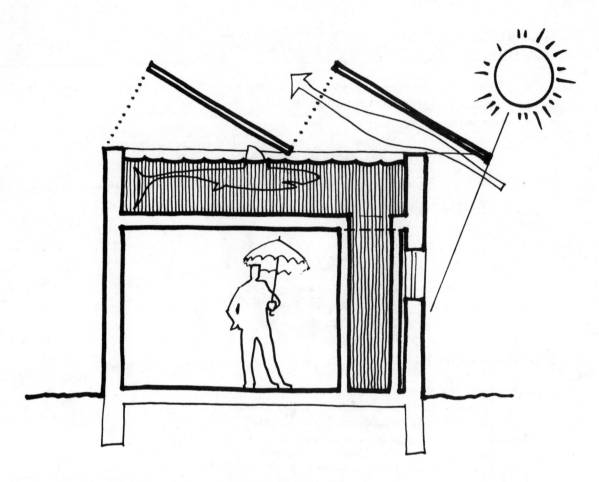

The "Cool Pool" concept is being developed and tested in Davis, Indio, and Sacramento, California, by Living Systems. The roof pond radiates its heat to the sky. The coolest water settles down into the radiant wall panel, cooling the house. The warmer water rises up into the pond where it radiates to the sky.

Ground Cooling

Since the ground is nearly always cooler than the air in the months when cooling is required, the more a house is in contact with the ground, the cooler it will be. Build your house below grade or into the side of a hill to obtain easy ground contact. You can also partially bank (berm) the earth around your house or even cover it. High levels of comfort and serene quiet usually accompany well-designed underground housing. Just be careful to insulate the building in a way and to a degree that's appropriate to your region. In cold climates, insulate the walls well from the ground. In mild climates, use less insulation. In warm climates, no insulation is needed; the earth will keep the house cool. In humid climates, provide ventilation so that the surfaces in contact with the ground are kept dry.

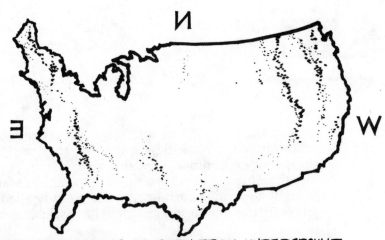

THE UNITED STATES AS SEEN FROM UNDERGROUND

REMEMBER, YOU SAW IT FIRST RIGHT HERE.

Earth Pipes

Buildings built near natural caves have long used underground air masses to provide ventilation, needing only a little heating in most seasons. Earth pipes for the same purpose are just starting to be used. Earth-pipe systems have been designed to use pipe ranging in diameter from 4 to 12 inches. Forty to as many as 200 feet of pipe have been buried 3 to 6 feet below the earth. Metal culverts and plastic and metal waste pipe have been used. As house air is vented to the outside, either naturally or with a fan, outside air is drawn through the pipes and then into the house. During the winter, the air is warmed. During the summer, the air is cooled; and in humid climates, moisture condenses out of the air and onto the surfaces of the tubes. The pipes are sloped slightly outward, away from the house, to carry away the moisture left behind by the humid air.

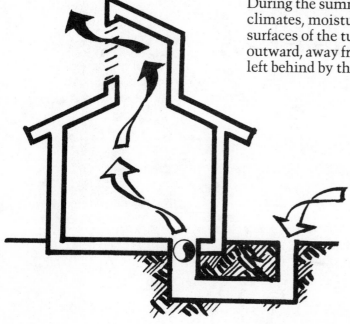

The earth surrounding such pipes has in many cases been warmed (or cooled) too quickly to the temperature of the incoming summer (or winter) air, causing the pipes to lose their intended usefulness. Considerable research is being done in this area, however, and more detailed design and construction information will be available soon.

PART 4

Putting the Pieces Together

Bricklayers of the 19th century using solar heat-storing masonry to build walls. (Woodcut. The Bettmann Archive.)

Chapter 9

House Design Based on Climate

"Will passive solar heating and cooling work where I live?" Simply put, the answer is "Yes!" After all, the sun shines everywhere. However, the question is not "whether" it will work, but rather, "how." What is the best way for you to use the many applications we have described in this book? The most important factor influencing the answer to this question is climate.

A Tradition of Regional Architecture

The amount of time the sun shines differs from one part of the country to another. Temperature variations are even greater. Wind conditions vary from calm to continuously windy. Some areas are arid, while others are humid.

It really wasn't so long ago that we used rather primitive methods of heating and cooling our homes: fireplaces in the winter and windows in the summer. We had limited access to materials to improve the thermal performance of our houses. We had to rely on local building materials such as logs, stone, adobe, or sod. Glass was scarce, window screen non-existent, and sawdust substituted for insulation.

Combined with the cultural diversity of the immigrating settlers, the forces of climate and limited building materials resulted in a rich variety of architecture fom one region to another.

For example, the harsh winters and abundant forests of the North Country combined to give us the traditional saltbox. Its compact floor design is oriented to the south

101

and is covered by a long north roof to shield it from the cold winter winds. In the Great Plains, sod substituted for wood, and subterranean shelters offered protection from harsh, winter westerlies. The oppressive humidity of the mid-Atlantic and Gulf states gave us narrow floor plans, floor-to-ceiling windows and doors, high ceilings, and broad overhangs for enhancing summer ventilation. In the Southwest, Spanish settlers used thick adobe walls and shaded courtyards with fountains to keep interiors cool during the summer and warm during the winter.

Simplified heating and cooling technology developed more quickly than improved materials and techniques for upgrading the thermal performance of houses, in part because of abundant and cheap energy. The result is that large central heating and cooling systems run by cheap energy compensate for climatically inappropriate house designs. For example, torrid summer solar gains through large west-facing windows in the arid Southwest have become as commonplace as horrendous heat loss through large north-facing single-layered windows in the brutally cold North. Building materials are easily transported from one part of the country to another at great energy expense. Petroleum based plastics imitate adobe in the north and window shutters in the south. Monotonous-looking subdivisions have become Anywhere, USA.

Now the era of unlimited cheap energy has passed. We again have the opportuniy to design houses that work with the climate and not against it. To make best use of this oportunity, we must understand the wide variety of energy conservation and passive solar heating and cooling applications, so as to appropriately select a suitable combination for a particular climate. In doing so, we will obtain the highest possible comfort at the lowest possible expenditure for materials and energy.

A Revival of Climate-Based Architecture

Climate, then, should have a major affect on your selection of energy conservation options and passive solar heating and cooling features, and, in turn, the architectural design of your house. The relationship between these factors and climate is discussed here to assist you with your choices. You will find that there are enough possibilities to offer the potential of near-zero heating and cooling bills nearly everywhere in the country while still satisfying your other needs for a home.

For simplicity, we can categorize climate into three types:

1. winter-dominated climates,
2. summer-dominated climates, and
3. climates that have a relatively balanced mix of winter and summer.

No doubt, you know which climate type you live in without having to look at a map of the country divided into the three zones. Besides, hills, valleys, lakes and rivers can cause your micro-climate to differ significantly from what a map would tell you.

Winter-Dominated Climates

Northern New England, New York, the upper Midwest, and most of the Rocky Mountain highlands are typical of cold, winter-dominated climates. Make your house compact in shape and consider earth-sheltering if your soil and terrain is not too rough. Pay attention to construction details for minimizing air-leakage. Consider a heat reclaimer for introducing fresh air preheated by exhaust air. Use 10 to 15 inches of insulation everywhere. Equip entrances with airlock vestibules.

Locate most windows as close to due south as possible, but certainly away from cold winds, which usually prevail from north and west. If summers are mild, moderate-sized east- and west-facing windows will rarely create overheating problems. Avoid east- and west-facing glass where summers get hot. Alternatively, shade this glass well and, especially in humid climates such as New York, Pennsylvania and the upper Midwest, use them to admit cooling breezes. Roof overhangs can shade south glass and help trap gentle air currents for additional ventilation. Shade trees are particularly effective as "air conditioners" in winter-dominated climates that also have hot summers.

Many winter-dominated climates, such as much of the Rockies, experience mild summers, and cooling need not be a major consideration when designing a house to minimize winter fuel bills. Some parts of the Rockies, however, are hot during the day and cool at night. Therefore, thermal mass, important in this region in the winter for storing plentiful sunlight, can effectively store evening coolness to keep houses comfortable during the day. Supplementary evaporative cooling from pools, fountains, or "swamp" coolers may be needed in climates with hot, arid summers, such as the Great Plains and the plateaus of the Rockies.

In winter climates with abundant sunshine, such as Colorado and the Southwest, passive systems can be sized large enough to economically provide more than 75 percent of the heat. The greater the percentage of the heat

supplied by the sun, the more thermal mass is required for storing the excess heat for successive cloudy days. Thus, the most effective systems are solar windows in combination with concrete, stone, brick, or adobe construction and solar walls. The same mass keeps these houses cool in the summer.

In most of the rest of the country, the sun shines an average amount. Therefore, the choice of passive systems should be based on factors other than the availability of sun. For example, a greenhouse may be a good choice if additional moisture is desired in the house, but it may be a poor choice in a humid climate. On the other hand, a sunporch in a humid climate can be converted to a screened summer porch.

The thermal mass of large solar window systems and solar walls is not as appropriate in winter-dominated climates that are also mild. In the Pacific Northwest, for example, windows and solar rooms with wood-framed houses will suffice without thermal mass. The windows can be used for summer ventilation.

Because passive systems lose more heat in cold weather than in mild, windows and solar walls should have double glazing and movable insulation; triple glazing should be used in *really* cold climates, 9000 degree days or more. Solar chimneys should be double-glazed in really cold climates. They are relatively more cost-effective than other passive systems in colder climates. Solar rooms produce less heat and cost more in cold climates. There may be other reasons for building them, however.

The extra glazing increases the cost of a cold-climate passive system. And, since conservation saves more energy in cold climates than in mild ones, conservation can be carried "to extremes," and the passive system can be limited in size to 10 or 20 percent of the floor area. Active solar systems, too, can be considered seriously because long heating seasons increase their cost-effectiveness. Solar water heating, of course, should be used regardless of climate. And be sure to slope your roof southward for future conversion to solar electric cells or to additional solar heating.

Summer-Dominated Climates

At the opposite end of the thermal spectrum are climates dominated by summer. Some parts of the country, such as Hawaii, Southern California, and Florida, have no winter. Light, open construction permits houses to capture all available air movement. With proper shading, these houses should require no mechanical air conditioning other than occasional fans in humid climates and pools or fountains in arid ones.

In summer-dominated climates, winters (if they exist at all) are usually quite mild. People tend to design houses that can be buttoned up easily when temperatures drop. A design dilemma then results, especially in humid climates such as the Gulf states. A house may be unable to be opened widely enough to permit sufficient cooling from ventilation. However, the same house may not be tight enough for efficient conventional heating and cooling.

A partial solution to this dilemma is a house that is both well insulated and well shaded. Windows should be double-glazed. For removing the winter chill, one third to one half of the total window area should face south. Properly designed houses will require little if any other heat in most winter climates of less than 3000 degree days per year.

The rest of the windows should be strategically placed to capture and enhance breezes. Locate large windows or vents as high as possible. Belvederes, turrets, gables, wind turbines and fans will add to the ventilating chimney effect. Place intake windows close to the floor and if possible, facing the winds. High ceilings help restrict warm air to levels above people's heads. Avoid shade trees in humid climates since they tend to stifle breezes, and their respiration adds moisture to air.

Make roofs light in color and vent all attic space well. Elevate houses slightly to assure good drainage and to minimize humidity. Carefully moisture-proof all floors and walls in direct contact with the ground but do not insulate unless cold, moist interior surfaces are considered undesirable. If you use earth pipes for cooling, make sure they are moisture-proof to avoid further humidification of incoming air by the ground. Since ground temperatures remain warm from the summer unless chilled by winter weather, this cooling method does not work in climates without winters. Even in climates with mild winters, the pipes must be buried more than six feet deep to obtain sufficiently cool temperatures.

(WITH APOLOGIES TO THE MASTER)
—MW

In arid climates, earth contact can provide pleasant relief from hot weather and does not have associated moisture problems since humidity is often encouraged. Earth pipes can be perforated to increase humidification of the air by the ground.

Thermal mass is usually of little advantage in mild, humid climates. Solar windows and chimneys can accomplish most if not all of the heating. Solar chimneys can also be used to enhance summer ventilation. Solar walls are less applicable unless summer nights are mild, in which case the mass of the wall can help keep the house cool during the day. Solar rooms work best in humid climates if moisture respiration from plants is minimized. A good solar room in this climate is a sunporch which is easily converted to a screened porch during the summer.

In general, passive systems can be single-glazed. However, double glazing of windows can help eliminate the need for a heating system or, in less mild winters, reduce the size of the system to a single space heater.

The design strategy is somewhat different in arid, summer-dominated climates such as Southern California and the deserts of the Southwest. Conventional construction is of massive materials such as adobe. Tall, narrow east- and west-facing windows are recessed for easy shading. South walls of heavy materials need not be covered with glass or plastic in mild areas. They will provide adequate heat in the winter and can be shaded to reduce solar gains during the summer. East, west, and north walls can be insulated to reduce both winter heat loss and summer heat gain. The addition of a second layer of glazing to windows can eliminate fuel bills in many arid, summer-dominated climates. Extra south windows or a solar room may sometimes be needed to achieve this.

Cooling bills can also be eliminated by taking further steps. Pools, fountains, and other evaporative cooling techniques are ideal except where water is a scarce resource. Pool ponds and sprays work well, too. Sky radiational cooling is effective if night skies are clear, and solar roofs should be considered. Walls and roofs should be light in color to reflect summer heat.

A Balance Between Winter and Summer

In many parts of the country, neither winter nor summer dominate. Year round mild weather prevails in San Francisco and its environs, but in the Southern Midwest, both summer and winter are equally harsh. Southern New England is considered moderate.

First, the mild areas. These are the climates of greatest design freedom. Thick insulation, proper orientation, and a modest-sized passive system can all but eliminate heating bills. The solar roof house in Atascadero, California has proven that (see page 126).

Shading, thermal mass with night cooling, and ventilation can do most or all of the cooling. Sky radiational cooling using roof panels can also be used.

Moderate climates hold many of the same opportunities as mild climates; they offer numerous design choices. However, mistakes are not as forgiving so that more careful design is necessary. Elimination of backup heating and cooling is more difficult, but possible.

The best approach is to find local houses 50 to 100 years old that are comfortable both summer and winter. For your own design, retain their thermal properties and embellish them further with energy conservation and passive heating and cooling measures. Chances are, you will have a very successful, low-energy house.

Harsh climates may be either humid or arid. Review the earlier parts of this chapter, and combine heating and cooling measures that do not conflict with one another. Lots of insulation and multipaned windows, for example, rarely compete with other measures. It is the high fuel bills that are typical of these climates that offer both the substantial incentive and the substantial reward for using energy conservation and passive solar.

Chapter 10

The Hard Choices of Cost

How Large?

We are now prepared to determine the proper size of a passive system. "Size" almost always refers to square footage of all glazed surfaces facing the sun, within 30 degrees east or west of due south. Other dimensions, such as the thickness of a solar wall or the floor area of a greenhouse, are fairly standard, and we have already covered them. The size of a system is the net square footage of glazed surface after subtracting the framing and trim. A wall of solar windows 8 feet high and 20 feet long (160 square feet) may be as much as 40 percent framing and 60 percent glazing, leaving but 96 square feet of window to catch the sun. It is possible, but difficult, to minimize framing to 5 or 10 percent of the surface area.

"collector"

There are many ways to size a passive system. As with active systems several years ago, engineers and scientists are producing dozens of documents crammed full with charts, graphs, and tables for determining the precise optimal size and performance of passive systems. This detailed information has its uses, but do not feel obligated to use it or to rely on it. Remember that Nature is forgiving and does not require nearly the precision that our computers tell us we do! Instead, review again some of the earlier pages that discuss size, and you will have a good "feel" for the surface area you need.

The first question you should answer is this: How large do you *want* the system to be? Believe it or not, answering this question is by far the most common, direct, and useful method of determining size. Most people want their system to be as large as possible to save as much energy as possible. In general, this is an excellent approach to take, and you may want to determine how to cover the entire south side of your house with passive devices. Many people do, just as the size of the roof often determines the size of active solar heating collectors.

You need not go to this extent, of course. A super-insulated house can be 50 percent solar heated with only slightly oversized south windows. But if you desire to reduce fuel bills to a minimum, large systems are generally in order. The trick is to design them right so that you save as much energy as possible with the least expense, and so that you are pleased with the appearance and comfort of your home.

Generally speaking, an optimal size for a passive system is that which supplies the same portion of heat as the "average percent of the time that the sun shines during the winter." For example, if your heating season is October through April, determine your local average winter sunshine from the U.S. monthly sunshine maps in Appendix 3. This percentage, say 55 percent for much of the country, is the portion of your heating load that can be supplied by a passive system in a well-insulated house with good assurance that the design can look beautiful, perform well, and be cost effective.

This is not to say that you should not try for larger percentages to be supplied by solar. But the design gets trickier, so consider obtaining additional advice from a

---

Oil, Electricity, Gas

A gallon of oil would supply 135,000 to 140,000 Btus if your furnace burned at 100 percent efficiency. It doesn't. Most furnaces, in fact, burn at between 40 and 70 percent annual efficiency, supplying the house with between 56,000 and 100,000 Btus per gallon burned. A useful average to use is 70,000.

One kwh of electricity is equivalent to 3400 Btus. Most electric resistance heating is 100 percent efficient (although three to four times more energy is expended at a power plant than what finally enters your house as electricity). Thus, 20 kwhs (68,000 Btus) of electric resistance heating supplies the heat obtainable from one gallon of oil (70,000 Btus). A proper application of an electric heat pump can supply twice as much energy per kwh as resistance heating.

One cubic foot of gas contains 1000 Btus of heat. One hundred cubic feet (1 ccf) contain 100,000 Btus (1000 cubic feet is represented by 1 mcf). At 60 percent annual furnace efficiency, each ccf of gas supplies 60,000 Btus, close to that of one gallon of oil or 20 kwh of electric resistance heating. For comparison, 1 gallon of oil may cost $1.25, 20 kwh of electricity $1.20 (6¢/kwh), and 1 ccf of gas $0.50.

local "expert," a workshop, or more books. The resources at the back of the book can provide some helpful hints on where to look next.

The next step is to determine your annual fuel consumption in gallons of oil, kilowatt hours (kwh) of electricity, or hundreds of cubic feet of gas (ccf). For an existing house, look at fuel bills. For new houses, calculate heat loss (get help or consult books—it's easy!). It's not unusual for an old, rambling Victorian house during a cold winter to use several thousand gallons of oil, 40,000 kwhs of electricity, or several hundred thousand cubic feet of gas. But a compact, super-insulated house may use as little as one fifth this amount—before the addition of passive.

The next step is to determine the energy output of the system. This is usually considered the hardest step, requiring the charts and graphs we mentioned earlier. However, it need not be difficult. In fact, it can be relatively easy.

Each square foot of a design that is appropriate for the climate (for example, a design that has the proper number of glazing layers), will supply the heat obtainable from roughly one gallon of oil (60,000 Btus). In cloudy climates, such as Buffalo, the amount can be half that. In the sunny Southwest, the output can double.

Thus, a house with an annual heat loss of 500 gallons in a climate of average (50%) sunshine will require 250 square feet of passive system to cut its fuel bill in half.

When selecting among the various passive applications, review the previous chapter for their appropriateness to your climate. In general, start with windows as the basic system and add other systems as needed or desired. In the previous example, the home may be designed for 100 square feet of south windows.* They will replace 100 gallons of oil or 20% of the total 500 gallon load. As discussed in earlier chapters, no special effort needs to be taken to add thermal mass since conventional construction usually provides adequate storage for solar windows, which supply 20 percent or less of the heat.

The next 150 square feet, supplying the next 50 percent of the load, might, for example, consist of a greenhouse, a solar wall, or a combination of the two. Remember to take into account the trim and framing when determining square footage.

*

What Does Passive Cost?

It seems that many people are willing to spend far more money on solar than on energy conservation (buying *more* energy rather than using *less!*). Conservation and solar are both important, and both can reduce your fuel bills enormously. But if you are willing to spend $3,000 on solar to save $200 of fuel per year, you should be willing to spend $2,000 on energy conservation measures to annually save $200 of fuel. Conversely, if you spent $1,000 for insulation and saved $200 last year on fuel, then think about spending $1,000 on solar to save $200 more next year.

Don't be fooled by the glamour of solar energy. Energy conservation, perhaps less glamorous, can go a long way toward solving our energy problems. We can economically super-insulate homes so that they use as little as 20 percent of the energy that conventional homes use, without significant use of solar energy. In moderately cold climates, walls and roofs should have R-values of at least 25 and 40, respectively. Use triple glazed windows, and caulk and weather strip. In cold climates, increase these R-values of the walls and roofs by 50 percent and add movable insulation over the windows. If you do anything less than these measures, you should restrict your expenditures on solar to $1,500. If you meet these measures, go all out with passive solar. However, you probably won't need to spend more than $5,000.

From a practical point of view, energy conservation measures are necessary in most of the U.S. climate to keep the required solar heat collection area small enough so that it can be easily and economically added to the average house. Properly-designed solar glazing totalling 20 to 40 percent of the floor area of your house (depending on the climate), will provide 50 to 80 percent of the heat. This is for a well-insulated house with adequate thermal mass for heat storage. (Super-insulated houses need less southern exposure to achieve this.) Thus, the collection glazing can cover much of the south-facing walls (or roof). Poorly insulated houses cannot normally obtain a large percentage of heat from the sun except in mild climates.

In cold, cloudy climates, energy conservation measures are usually more cost effective than are solar heating devices, at least until heating loads of a house are reduced by 40 to 60 percent compared with previous heating bills.

But the house is still a "conventional" one compared with super-insulated houses. In mild, sunny climates, solar heating and cooling may be a more cost effective means of reducing fuel bills than energy conservation since the climate permits passive systems to have fewer layers of glazing.

Passive system costs can vary enormously, depending on size, design, materials, and construction methods, and passive system type. Keep the sizes manageable. For example, a 15-foot-high wall of glass is harder to build than an 8-foot-high wall. Don't make the design complicated. *Keep it simple!* An occasional fan to help move air from one part of the house to another—for example, from the south side to the north or from upstairs to downstairs—can keep floor plans from becoming contorted and passive systems from becoming crude contrivances.

Use basic, easy-to-handle materials, preferably locally produced. Build the systems yourself, if possible. This can save 40 to 60 percent of the cost. Contractor-built systems are most cost effective on new housing rather than on existing. However, additions and major remodellings are excellent opportunities for improving the cost effectiveness of contractor-built systems for existing houses.

Which of the various passive system types you choose also affects cost. *Solar windows* cost nothing if what would have been a north window is instead placed facing south. Solar windows can also be inexpensive if you are willing to have some of them fixed in place, nonoperable. Fixed glass can be one third the cost of operable windows.

If you intend to build of heavy materials regardless of using solar, your thermal mass comes free. On the other hand, concrete floor slabs or brick partitions added to conventional wood-framed construction can add to the cost.

In mild climates, the windows need have only two layers. In severe climates, triple glazing and movable insulation are usually necessary (and cost effective!). Shades or awnings may add to the cost in southern climates.

Solar windows, then, can cost nothing or as much as $20 per square foot.

If you are able to easily convert your south wall to a *solar chimney* by simply adding a single layer of glazing, the materials will cost but a few dollars per square foot. If you buy a solar collector and hire someone to install it, the final cost could well be three to four times more. Either way, if the collectors are large enough to provide only 15 to 20 percent of the heat, no new thermal mass will be needed. The cost of making your south wall into a solar chimney, then, will be primarily that of just the collector or glazing. When more than 20 percent heating is obtained, however, thermal mass must be added to accommodate the heat. This makes solar chimneys a whole new ball game, and, of course, costs increase.

The cost of building a *solar wall* is less when the south wall would have been built of concrete block, brick, stone, or adobe anyway. Keep in mind that stronger foundations are needed to support the weight and that extra floor area may be needed to accommodate the thicker-than-normal wall. In mild climates, only single glazing may be needed, while in cold climates, as many as three layers of glass or movable insulation may be necessary. Thus, solar walls may cost only a few dollars per square foot for the glazing or up to $25 per square foot.

Cost estimates for *solar roofs* are more elusive. A strong support structure, the water ponds, and the movable insulating covers can be costly. On the other hand, Harold Hay's "skytherm" house in Atascadero, California, if built in large numbers, might cost no more than conventional housing since, in that climate, it requires no back up heating or cooling systems. According to Mr. Hay, these cost savings completely pay for the passive system. In fact, the conventional heating and cooling system can be greatly reduced in size and complexity, or can be eliminated entirely from many energy-conserving, passive solar homes throughout much of the country, saving several thousand dollars in first cost and hundreds of dollars in maintenance costs. Small space heaters or wood burning stoves often suffice in the winter, and fans or modest-sized window air conditioners or evaporative coolers do the job in the summer.

A single glazed *solar room* can cost only a few dollars if the frame is built of used lumber and covered with a thin sheet of plastic. Commercially sold greenhouses, however, can cost thousands of dollars. Solar rooms reduce the cost of other building components. For example, when protected by a solar room, walls that would otherwise be exposed to the weather can be simplified and reduced in cost.

Whatever system or combination of systems you choose, "shop around" fully. Seek advice and look for the best prices. Pay attention to detail and use passive systems to enhance the beauty of your house.

Will you be satisfied with the fuel savings resulting from your expenditures for passive systems? The easy answer is "Yes!" However, everyone's expectations and needs are different. For example, some people require that passive pay for itself in five years (say a fuel savings of $100 per year on a first cost of $500). Others can afford to be more far-sighted and can view conservation and solar costs as investments in the future. Such people realize that short-sightedness got us into our energy mess, and longer term investments of 15 to 25 years can help get us out. Keep in mind that energy prices keep doubling every few years, and as they do, your "minimum energy dwelling" is perceived as more and more valuable by your friends, neighbors, . . . and the next buyer.

As a matter of fact, you can disregard the energy savings from many passive systems in determining whether they are good investments. Instead, they can be viewed as an integral part of your house, increasing in value along with the whole house. Thus, when the value of a house inflates by 50 per cent in say, five years, so will its solar greenhouse! If the house is then sold, the owner receives back his entire investment in the greenhouse and then some . . . plus, five years of lower energy bills, too.

In this sense, solar energy is indeed free. But passive solar is also convincing on the basis of dollars spent for the energy saved.

A conventional house may use $1,500 of fuel per year, and perhaps twice that in a few years as energy prices rise. For the many solar and conservation measures recommended in this book, an expenditure up to $10,000 may be necessary, depending on climate, the size of the house, and which conservation and solar features are selected. The result will be fuel bills reduced by 70 to 90 per cent, or $1,350 per year, and as much as $2,700 when prices double. Even if you must spend $10,000—and most people will spend less than $7,000—your annual mortgage payment will increase about $1,000 ($85 monthly) for the solar/conservation investment, but the energy savings more than compensates. Federal energy tax credits of 15 per cent for conservation measures and up to 40 per cent for solar, will offset the extra down payment you need as a result of the somewhat higher cost of the low energy house due to its energy-saving measures.

Thus, your cash flow is positive and the annual cost of owning an energy-conserving, passive solar home is less than owning a conventional one.

Afterword

You have now reached the end of the book and are at least 7 percent smarter than you were when you started it. Now it's time to put all your new-found knowledge to work. The things you've learned are of great value for they can affect not only *your* pocketbook, *your* comfort, and *your* way of life, but global politics, environmental quality, and the value of the dollar as well. We are in a great period of transition—from the fossil-fuel age to the solar age. Standing right up close to that transition, it's sometimes hard to sense the magnitude of the changes now taking place. But each day brings a bit more perspective to our view of the revolution. Each day makes solar energy systems look better and the nuclear nightmare look worse. Each step we take in the direction of a solar civilization will have untold benefits to us, to our children, and to all the generations to come, not to mention the benefits to all the other creatures of the world, whose fate is so inextricably tied, now, to ours.

Photo Section

The houses in the photographs on the following pages are representative of the thousands of passive solar homes being built or retrofitted throughout the country. The variety of approaches is truly astounding; proof that solar offers new opportunities for innovation, creativity, and beauty.

Solar Windows

Solar windows can be built into the south side of any new home. This energy-efficient, wood-framed house has almost no windows facing north. Leaf-shedding trees allow sunlight inside during winter but provide the house with shading in summer. The photo below shows the winter sun warming the south-facing bay of the house. A centrally placed woodstove provides the necessary back-up heat. Its fuel is also a form of solar energy. (Designed by Elizabeth Cohn and Total Environment Action, Inc., especially Bruce Anderson and Doug Mahone; built by Ned Eldredge.)

Total Environmental Action, Inc., Harrisville, NH

TEA, Inc., Harrisville, NH

Alan Ross, Milton, MA

Throughout the long snowy Colorado winter, south-facing windows collect solar heat to warm this house, one of 150 solar homes in the Aspen area. Thermal insulating blinds trap heat inside during winter nights and are used as shades on summer days to help keep the house cool. Note the frame of the greenhouse under construction. (Designed by Alan Ross, built by Tony Fusaro)

The stairwell of this house in Minneapolis was transformed into a two-story sunspace through the addition of tall bay windows. Other solar windows were added to the south wall as well, but the first steps were to heavily insulate the attic and walls and to take other energy conserving measures. Phase change heat storage rods (see page 9) are used in place of ordinary thermal mass. (Designed by Peter Pfister, construction by the Architectural Alliance.)

Solar Windows

Patio door glass was used in this low-cost window retrofit of a south-facing sunporch. This is a common and effective way of converting existing homes to passive solar heating. (Designed for New England SUEDE, construction by the Center for Ecological Technology. See page 129 for construction details and additional credits.)

Architectural Alliance, Minneapolis, MN

Center for Ecological Technology, Pittsfield, MA

Malcolm Katz, East Sullivan, NH

This New Mexico house incorporates solar windows across most of its south side. Thick adobe walls store solar heat. In summer, walls release heat at night, keeping the house cool during the day. Built-in overhangs shade the windows from the high summer sun. (Designed by David Wright, built by Karen Terry.)

Solar Chimneys

Solar chimneys can be easily included in new home construction. Framed panels with an absorber plate and glazing are mounted on the south wall, and solar heated air vents naturally upwards and into the house. Cool air returns through bottom collector vents to be reheated. In summer, vents can be closed to prevent unnecessary daytime heating. (Designed/engineered by TEA, Inc., especially Dan Scully/ Charles Michal, Jeremy Coleman; built by Charles Joslin.)

CET, Pittsfield, MA

Solar chimneys in the form of thermosiphoning air panels, or TAPs, were added to this home over the insulated south wall. TAPs are simple and inexpensive to construct and install, and provide heat to the house without fans or controls. (Designed for New England SUEDE, construction by CET. See page 131 for construction details and additional credits.)

The solar chimney on the west wall of this Denver house takes the place of a fan for cooling. Heated air rises in the panel and vents to the outdoors, pulling fresh air through the house. A wind-driven ventilator over the solar chimney also cools the house. Heavy insulation and a reflective blind in the east window help to cut cooling requirements. The roof collectors on this house are part of an active solar heating system. (Designed by D. J. Frey, Crowther Solar Group, built by Shaw Construction Co.)

Solar Walls

Floor-to-ceiling tubes made of Kalwall Sunlite™ are filled with water and placed in a row behind south-facing windows to provide heat storage for this small office in New Mexico. (Office of Richard Grenfell, AIA.)

Kalwall Corporation, Manchester, NH

Robert Perron, New York

This solar home in Connecticut uses several passive systems. Concrete walls and floors store heat from the solar windows on the left. The glazing on the right traps heat over a concrete solar wall. To the lower left is a solar greenhouse. Heavy insulation, roof overhangs and awnings, and heat-venting skylights help keep the house cool in summer. In addition, the mass absorbs excess summer heat and releases it when nights are cool. (Solar and house design and construction by Stephen Lasar.)

Paul Pietz, TEA, Inc., Harrisville, NH

This energy-efficient demonstration house at Brookhaven National Laboratory on Long Island uses a triple-glazed solar wall as one of its many passive solar features. Also included are solar windows, a two-story greenhouse, heavy insulation, and thermal mass for heat storage. In summer, heavy insulation and movable awnings keep out much unwanted heat and the mass stores any excess. Windows can be opened for ventilation, and no air conditioning is needed. (Designed/engineered by TEA, Inc., especially Paul Pietz, Lisa Heschong/Dan Lewis; built by Thermal Comfort, Inc. See page 134 for construction details and additional credits.)

This midwestern duplex was retrofitted for energy conservation and passive solar heating to turn the existing brick walls into energy savers. North, east, and west walls were covered with rigid foam insulation board and stucco to reduce heat loss. The upstairs south window was enlarged, and glazing added over the brick wall around it to create both a solar wall and a solar room. (Design and construction by Londe-Parker-Michels, especially Tim Michels.)

In this Davis, California house, water-filled metal culverts stand behind the two-story window to absorb solar heat. Water tubes are also placed in the greenhouse to pre-heat the household water. Symmetrical active domestic hot water collectors flank the second story window on the left. Well-insulated walls and ceilings, heat-venting eaves and skylights, roll-down shades, and a trellis with leaf-shedding vines all provide summer cooling. Operable vents at the bottom of the windows allow cool night air to enter. (Designed by Living Systems, especially Gregory Acker and Marshall B. Hunt; built by M. B. Hunt and Virginia Thigpen.)

These corrugated water tanks are galvanized steel culverts painted dark brown. They are 14 feet tall and have an 18 inch diameter. Six-inch spacing between the tanks permits sunlight to enter the house. (Living Systems.)

123

Courtesy, Brick House Publishing

Solar Rooms

This house shows another simple adaptation of a south-facing sunporch into an attractive and useful solar room. (Designed for New England SUEDE, construction by CET.)

CET, Pittsfield, MA

Solar rooms can be built as greenhouses for year-round food production. This solar greenhouse typifies the way in which solar rooms can be well integrated into the original house design. (Designed for New England SUEDE, construction by Southern New Hampshire Services. See page 137 for construction details and additional credits.)

Jennifer Harris, TEA Foundation, Harrisville, NH

Once abandoned, this St. Louis inner-city brick-walled building has been converted into a two-family home with a three story solarium. (Designed by Warren L. Cargal, Solar Building Corporation.)

Warren L. Cargal, Solar Building Corporation, St. Louis, MO

The large solar room on this house contributes heat to a heavily insulated envelope air space between the house and an outer shell surrounding the house. Warm air rises from the sunspace into the envelope and flows over the top floor ceiling, down behind the back north wall, through the crawl space under the floor and returns to the sunspace. Gable end vents under the east and west roof peaks ventilate excess summer heat from the sunspace. (House design and solar envelope by Ekosea, especially Lee Porter Butler; built by Madeiros Brothers Construction and Robert Mastin.)

An extra, warm, sunny daytime room has been added to this home by glazing in the south-facing sunporch. Another solar feature is the existing brick wall at the back of the room for heat storage. Heat conducts through the wall and radiates into the house. (Design and construction by Evog Associates, Inc.)

A two-story double-glazed solar room supplies heat to this New Mexico house. Adobe floors and walls provide heat storage, and also help to keep the house cool during the summer. The interior of this solar room is shown on the cover. (Designed by Sun Mountain Design, construction and solar engineering by Communico, especially Wayne Nichols.)

Solar Roofs

Water-filled, black, polyvinyl bags covering the roof absorb and store sufficient solar energy to meet 100% of the heating needs of this Atascadero, California home. Twenty-six tons of water rest on a metal pan ceiling. Rigid foam insulating panels slide over the roof at night. In summer, the panels cover the roof ponds during the day and are opened at night to allow the ponds to radiate heat to the cool sky, providing 100% of the cooling. (Skytherm concept and engineering by Harold Hay; house design by Kenneth L. Haggard and John Edmisten.)

These water-filled bags contain 2,000 gallons of water, weighing 8 tons. They are roughly 6 by 7 feet and 10 inches high, and rest in wood-framed supports across the attic floor. (Peter Hollander, E-M Architects.)

The attic space in this New England farmhouse was renovated as both collector and storage area. Kalwall glazing admits solar heat which is stored in black, water-filled bags. A fan distributes heated attic air throughout the house. A beadwall® system turns the glazing into night insulation and in summer blocks daytime solar gain. (Designed by E-M Architects, especially B. C. Ellis; built by Community Builders.)

Construction Details

This section features construction details from some of the more typical examples of the passive systems we have described. This information should give designers, builders, and homeowners a much better knowledge of how each system is designed and built, with a focus on those details needing special attention. Since full sets of construction drawings are not included here, you may wish to seek professional assistance before actually building your own system.

The details shown here are from four designs developed in two federally-sponsored demonstration projects to promote solar design, research, and construction. They include a **solar window**, **solar chimney**, and a **solar room** from Project SUEDE, and a **solar wall** from the Brookhaven House.

Project SUEDE, "Solar Utilization, Economic Development, and Employment," was part of a nationwide effort to train solar installers and to build solar applications into existing houses. Sponsored by the U.S. Community Services Administration, Department of Energy, and Department of Labor, SUEDE was carried out in New England by a four-member consortium: the Center for Ecological Technology in Pittsfield, Massachusetts; the Cooperative Extension Service of the University of Massachusetts in Amherst; Southern New Hampshire Services in Manchester, New Hampshire; and Total Environmental Action Foundation in Harrisville, New Hampshire. Together, these groups trained 30 installers who built one of three types of low-cost solar systems onto nearly 100 New England homes. A major goal in Project SUEDE was to demonstrate that solar designs can be simple, can be built at reasonable costs from readily-available building materials, and can be attractive and work well.

Examples of the New England SUEDE systems illustrated here also appear in the color section. The added solar windows, the thermo-siphoning air panel retrofit were each constructed by the Center for Ecological Technology. The attached greenhouse was constructed by Southern New Hampshire Services. Design for New England SUEDE were developed by Total Environmental Action, Inc., (TEA), of Harrisville, New Hampshire.

127

The Brookhaven House is the result of a research and design effort carried out by TEA, Inc., under contract to Brookhaven National Laboratory, and built at the Lab site on Long Island as a demonstration house to be monitored for its performance. The work was sponsored by the Building Division of the Office of Buildings and Community Systems, Office of the Assistant Secretary of Conservation and Solar Applications, U.S. Department of Energy.

The goal of the Brookhaven project was to develop an attractive, energy-conserving, single-family home of conventional design, using thermal storage materials in combination with heavy insulation and passive solar systems to significantly cut heating costs without reducing comfort. The construction details shown here are from the triple-glazed storage wall located next to a large sunspace and serving as the structural south wall of the dining room. This storage wall also contains a set of windows for direct gain, natural lighting, and a view from inside.

A photograph of the Brookhaven House appears in the color section.

The drawings here were prepared by and adapted by the authors from Total Environmental Action, Inc. designs for the Brookhaven and SUEDE projects. As neither the authors, publisher, TEA nor any of its employees, nor any of the original SUEDE and Brookhaven project participants, have any control over the final use of these revised drawings, all warrantees, expressed or implied, for the usefulness of these drawings and all liabilities which may result from the use of these drawings are voided by their use in construction.

It is good practice to have all dimensions, quantities, and specifications reviewed by a competent local architect, engineer, and/or building official prior to construction to assure compliance with individual requirements, and local codes and conditions.

Solar Windows

These details were developed for a low-cost addition of direct gain south glazing in standard 2×4 stud wall construction. A section of the south wall is removed and new framing added as shown to prepare for the addition of standard-sized insulated glass units. These fixed units are installed using standard glazing techniques including setting blocks, glazing tape and weep holes for condensation. The rough framing is finished with trim pieces and glazing stops. Note that cutting into the framing of a stud wall house can be a major structural alteration to the house, and should only be undertaken after professional verification that the new structure is adequate and that existing floor and roof loads can be carried safely during and after the renovation project. (Construction details, New England SUEDE.)

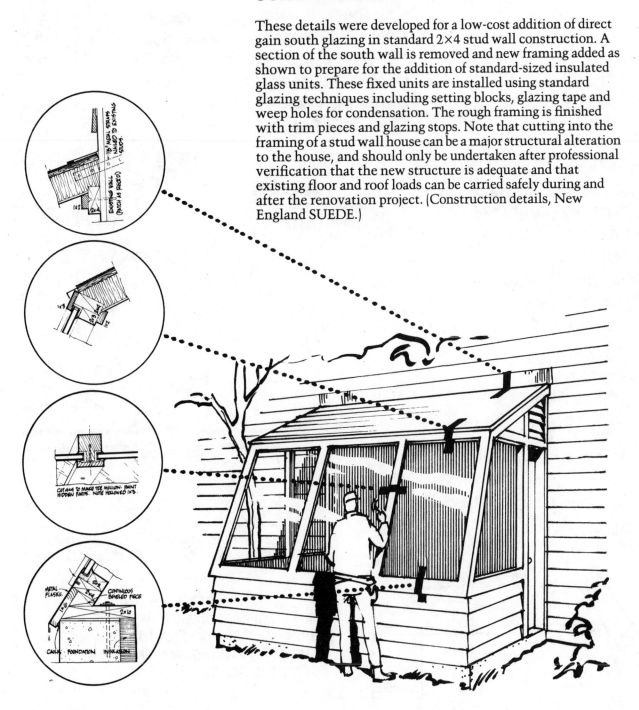

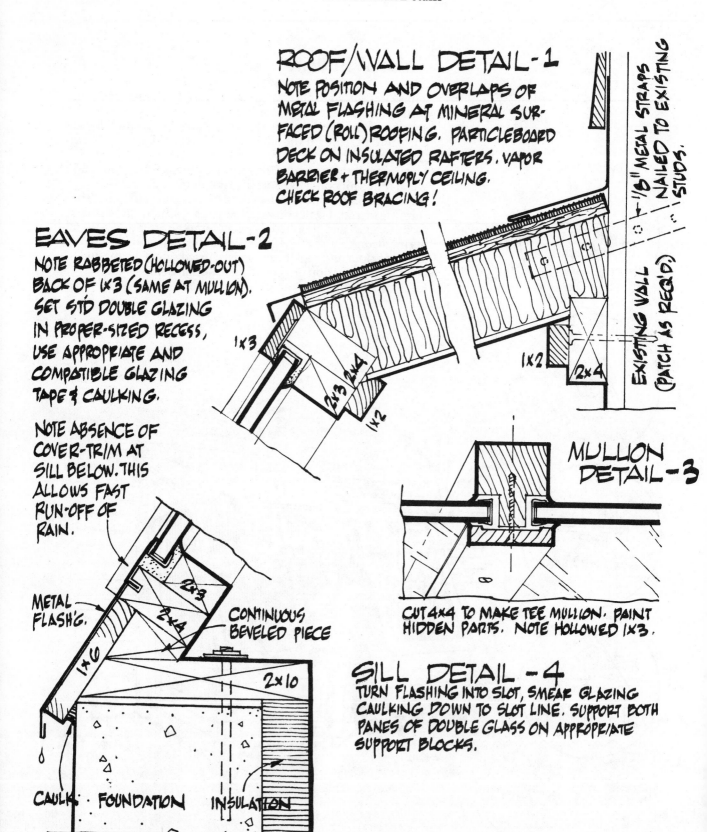

ROOF/WALL DETAIL-1
NOTE POSITION AND OVERLAPS OF METAL FLASHING AT MINERAL SURFACED (ROLL) ROOFING. PARTICLEBOARD DECK ON INSULATED RAFTERS. VAPOR BARRIER + THERMOPLY CEILING. CHECK ROOF BRACING!

1/8" METAL STRAPS NAILED TO EXISTING STUDS.

EXISTING WALL (PATCH AS REQ'D.)

EAVES DETAIL-2
NOTE RABBETED (HOLLOWED-OUT) BACK OF 1×3 (SAME AT MULLION). SET STD DOUBLE GLAZING IN PROPER-SIZED RECESS, USE APPROPRIATE AND COMPATIBLE GLAZING TAPE & CAULKING.

NOTE ABSENCE OF COVER-TRIM AT SILL BELOW. THIS ALLOWS FAST RUN-OFF OF RAIN.

1×3   2×3 2×4   1×2   1×2   2×4

METAL FLASH'G.

CONTINUOUS BEVELED PIECE

1×3   2×4   1×6   2×10

CAULK · FOUNDATION   INSULATION

MULLION DETAIL-3
CUT 4×4 TO MAKE TEE MULLION. PAINT HIDDEN PARTS. NOTE HOLLOWED 1×3.

SILL DETAIL - 4
TURN FLASHING INTO SLOT, SMEAR GLAZING CAULKING DOWN TO SLOT LINE. SUPPORT BOTH PANES OF DOUBLE GLASS ON APPROPRIATE SUPPORT BLOCKS.

Solar Chimneys

This retrofit passive space heating device, called a thermosiphoning air panel (TAP), uses the existing house wall as the major structural element. The exterior finish is removed, new Thermoply™ structural sheathing added over the existing wall, and wood framing added to support the ribbed aluminum absorber plate (industrial siding material) and to support the field-installed insulated glass units. The system shown uses three patio door replacement units as the aperture, creating three areas of absorber plate, each of which requires a high and a low vent through the house wall to allow the thermosiphoning action to occur.(See pg. 57 for damper construction tips.) The weight of the added glazing is carried by brackets at the base of the panel to a continuous ledger strip bolted to the house wall. After flashing is added, the exterior siding materials are patched around the unit to complete the installation. (Construction details, New England SUEDE.)

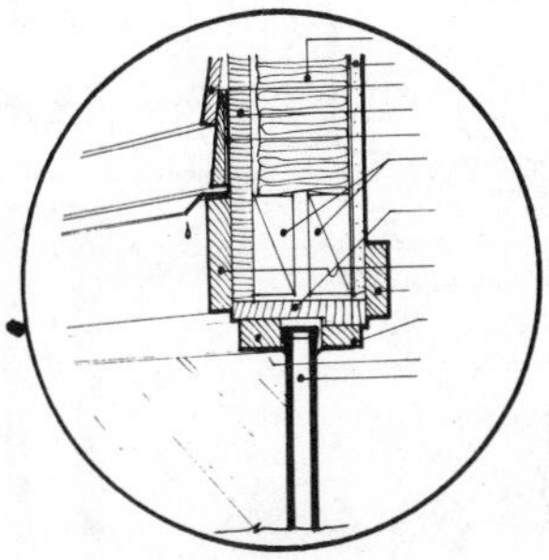

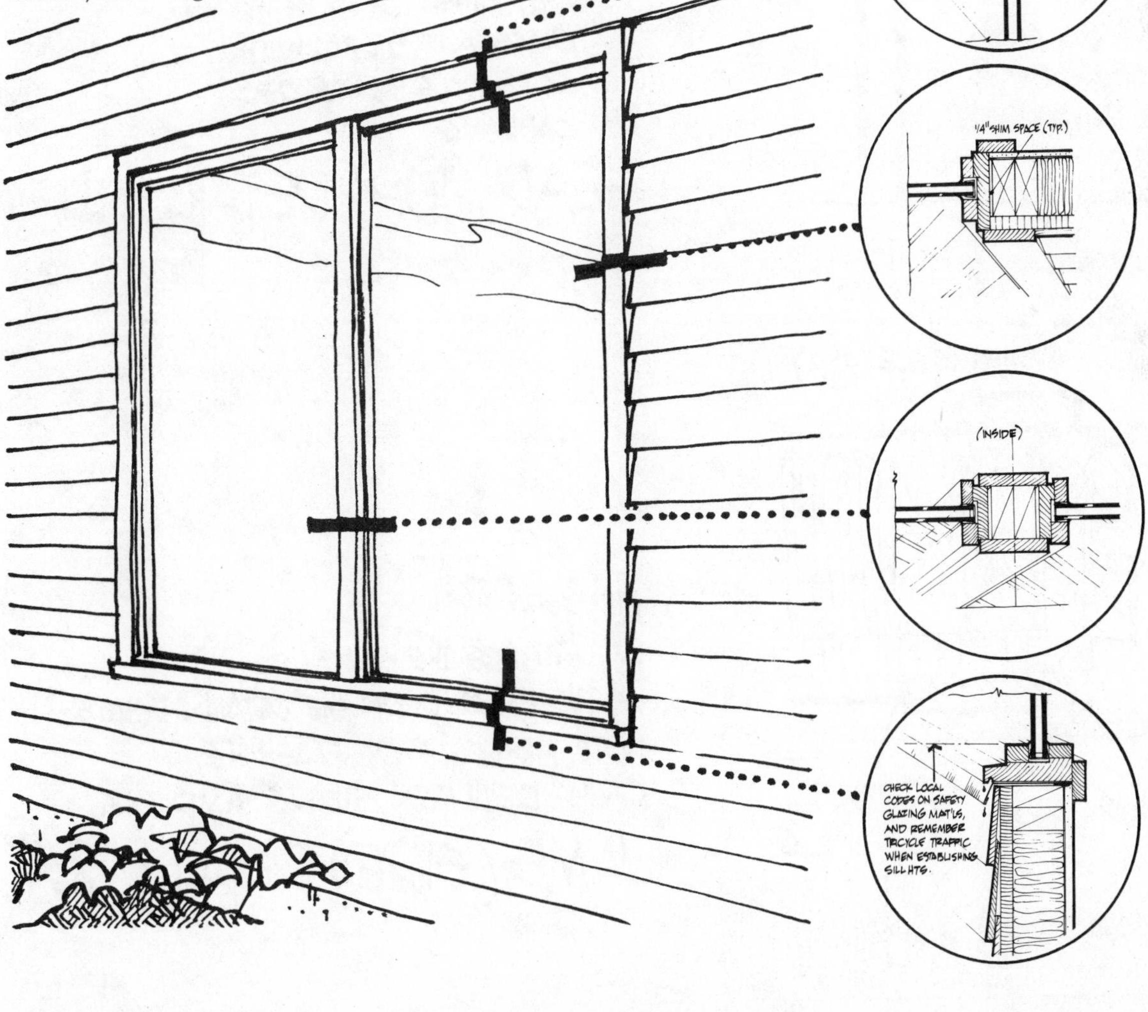

¼" SHIM SPACE (TYP.)

(INSIDE)

CHECK LOCAL CODES ON SAFETY GLAZING MAT'LS, AND REMEMBER TRICKLE TRAFFIC WHEN ESTABLISHING SILL HTS.

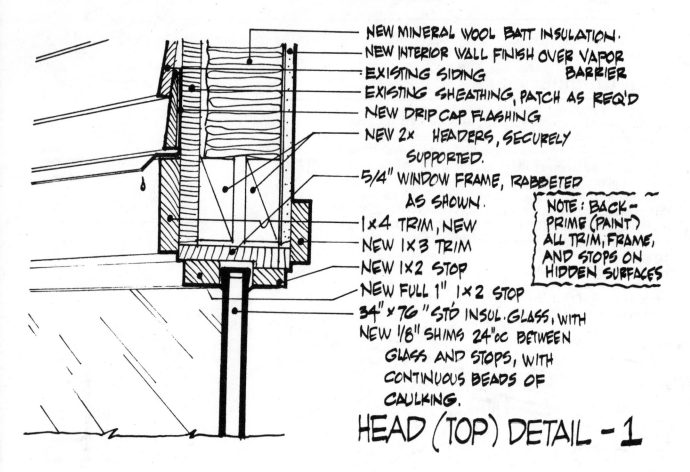

NEW MINERAL WOOL BATT INSULATION.
NEW INTERIOR WALL FINISH OVER VAPOR
EXISTING SIDING   BARRIER
EXISTING SHEATHING, PATCH AS REQ'D
NEW DRIP CAP FLASHING
NEW 2x HEADERS, SECURELY
        SUPPORTED.
5/4" WINDOW FRAME, RABBETED
        AS SHOWN.
1×4 TRIM, NEW
NEW 1×3 TRIM
NEW 1×2 STOP
NEW FULL 1" 1×2 STOP
34" × 76" STD INSUL. GLASS, WITH
NEW 1/8" SHIMS 24"OC BETWEEN
     GLASS AND STOPS, WITH
     CONTINUOUS BEADS OF
     CAULKING.

NOTE: BACK~
PRIME (PAINT)
ALL TRIM, FRAME,
AND STOPS ON
HIDDEN SURFACES

HEAD (TOP) DETAIL - 1

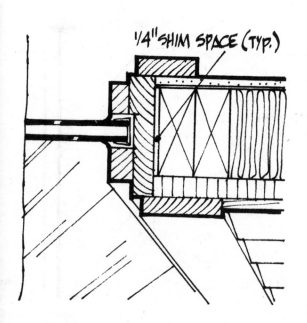

1/4" SHIM SPACE (TYP.)

MEMBERS SHOWN
AT LEFT ARE THE SAME
AS THOSE NOTED ON THE DETAIL ABOVE.
NOTE ABSENCE OF FLASHING.
CAULK BETWEEN SIDING AND 1×4.

JAMB (SIDE) DETAIL - 2

AGAIN, THE DETAIL IS SIMILAR TO THE HEAD DETAIL BUT NOTICE THAT THE 5/4" WINDOW FRAME IS 3-1/2" TO MATCH THE 2×4S, AND THE COVER TRIM, INSIDE AND OUT, IS 4-1/2" WIDE

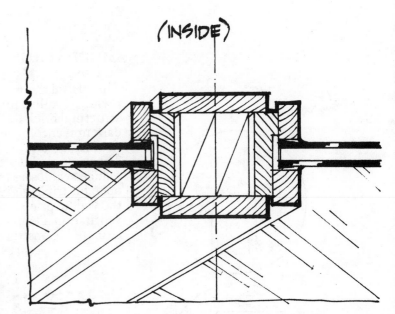

(INSIDE)

MULLION (POST) - 3

SILL DETAIL - 4

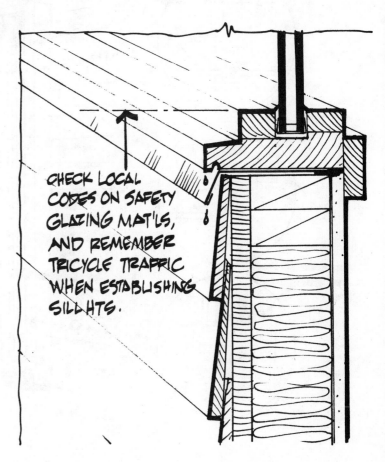

CHECK LOCAL CODES ON SAFETY GLAZING MAT'LS, AND REMEMBER TRICYCLE TRAFFIC WHEN ESTABLISHING SILL HTS.

NOTE THAT THE WINDOW FRAME AT THE SILL IS MADE OF 2X MATERIAL WITH TWO RABBETED STEPS ON TOP AND ONE DRIP SLOT ON THE BOTTOM.
CAULK THE UNDERSILL FLASHING NEAR THE INDOOR SIDE.

Solar Walls

This glazed thermal storage wall is comprised of glazing frame members milled from cedar 4×4's bolted to an eight-inch thick structural brick wall. The bricks are dense paving bricks—a dark umber color on the outside, standard terra cotta color on the inside—and are laid up with all cavities filled with mortar. The triple glazed panels, designed for use in the northeast, reduce heat losses to the outside from the warm wall. Standard operable triple-glazed casement windows are incorporated into the wall to provide direct gain heating, light, views and ventilation. Double glazing is suitable for use in milder climates. (Construction details, the Brookhaven House.)

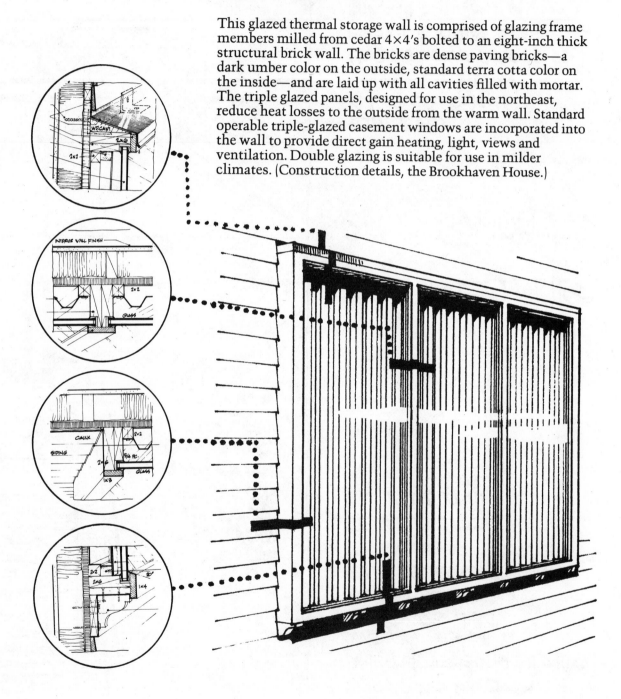

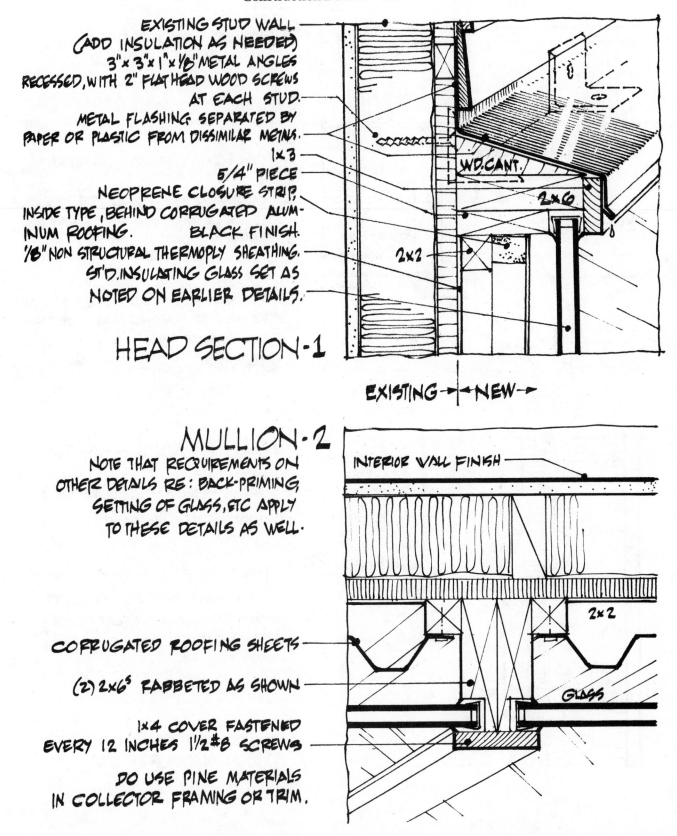

EXISTING STUD WALL
(ADD INSULATION AS NEEDED)
3"x 3"x 1"x 1/8" METAL ANGLES
RECESSED, WITH 2" FLATHEAD WOOD SCREWS
AT EACH STUD.
METAL FLASHING SEPARATED BY
PAPER OR PLASTIC FROM DISSIMILAR METALS.
1x3
5/4" PIECE
NEOPRENE CLOSURE STRIP,
INSIDE TYPE, BEHIND CORRUGATED ALUM-
INUM ROOFING.   BLACK FINISH.
1/8" NON STRUCTURAL THERMOPLY SHEATHING.
ST'D. INSULATING GLASS SET AS
NOTED ON EARLIER DETAILS.

WD. CANT.

2x6

2x2

HEAD SECTION · 1

EXISTING →|← NEW →

MULLION · 2

NOTE THAT REQUIREMENTS ON
OTHER DETAILS RE: BACK-PRIMING
SETTING OF GLASS, ETC APPLY
TO THESE DETAILS AS WELL.

INTERIOR WALL FINISH

2x2

CORRUGATED ROOFING SHEETS

(2) 2x6^S RABBETED AS SHOWN

GLASS

1x4 COVER FASTENED
EVERY 12 INCHES 1 1/2 #8 SCREWS

DO USE PINE MATERIALS
IN COLLECTOR FRAMING OR TRIM.

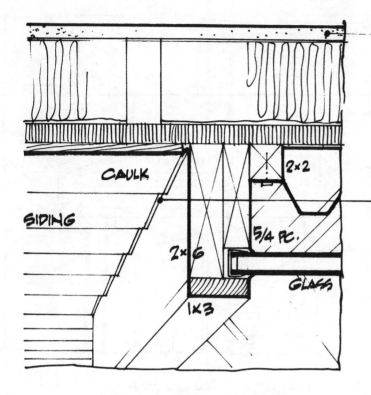

INTERIOR WALL FINISH.

SELECT 2×6 MEMBERS CAREFULLY!

LEAVE ¼" GAP AT SIDING FOR CAULKING;
USE MAT'L COMPATIBLE WITH FINISH ON
WOOD.

NOTE THAT VERTICAL PIECES ARE NOT
FASTENED TO EXISTING WALL; HORIZON-
TAL MEMBERS ARE.

JAMB DETAIL - 3

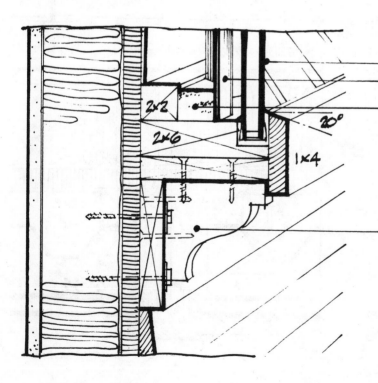

SILL DETAIL - 4

GLASS.

CORRUG. METAL AND CLOSURE STRIP.

NEOPRENE CLOSURE STRIP

SUPPORT BRACE: 2 CONTINUOUS 1×6 OR
5/4"×6" BOARDS WITH 2" BRACKETS
EVERY FOOT. LONG FLATHEAD SCREWS
FASTEN BRACE TO BRACKETS.

CONNECT BRACE TO WALL WITH
(2) 3" LAG SCREWS @ EA. STUD

Solar Rooms

This solar greenhouse uses stock size insulated glass patio door units as the solar aperture. These units are field-mounted in the wood-framed structure which rests on an added foundation wall of poured concrete or block and which is attached to the existing house wall by 2×4 braces and a 2×4 ledger strip bolted to the wall. The side wall can be either clapboard or other siding to match the house. In this design, the two-inch beadboard foundation insulation is located on the inside of the foundation wall to make a weatherproof exterior with no additional finishing required. An optional roll-down insulating curtain is included at the sloped glazing. (Construction details, New England SUEDE.)

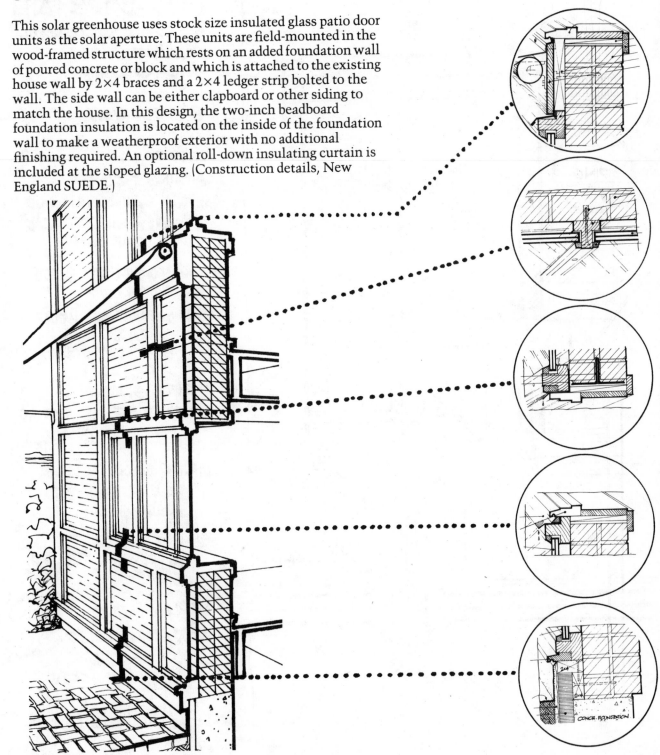

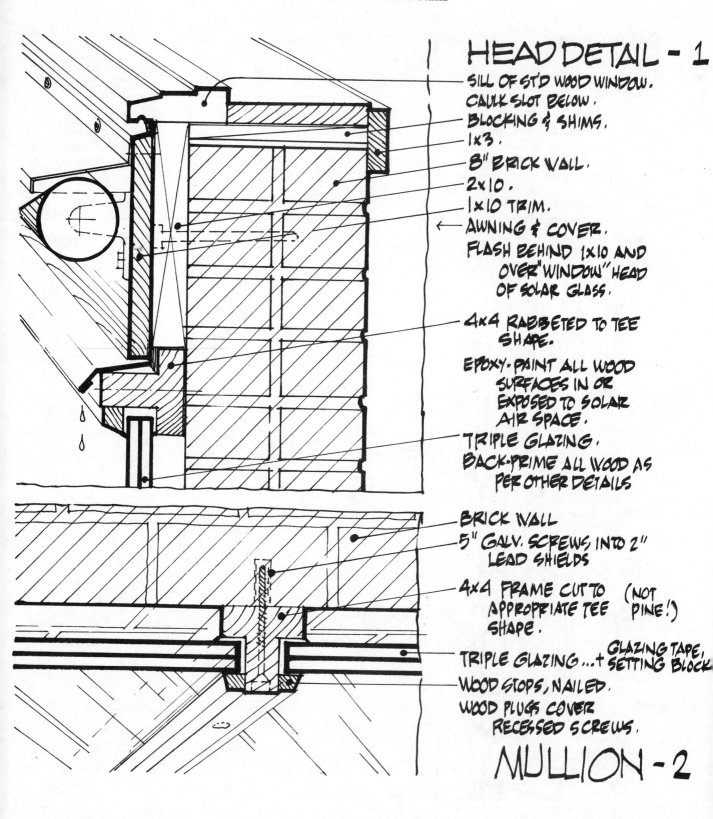

HEAD DETAIL - 1

SILL OF ST'D WOOD WINDOW.
CAULK SLOT BELOW.
BLOCKING & SHIMS.
1x3.
8" BRICK WALL.
2x10.
1x10 TRIM.
← AWNING & COVER.
FLASH BEHIND 1x10 AND
    OVER "WINDOW" HEAD
    OF SOLAR GLASS.

4x4 RABBETED TO TEE
    SHAPE.

EPOXY-PAINT ALL WOOD
    SURFACES IN OR
    EXPOSED TO SOLAR
    AIR SPACE.

TRIPLE GLAZING.
BACK-PRIME ALL WOOD AS
    PER OTHER DETAILS

BRICK WALL

5" GALV. SCREWS INTO 2"
    LEAD SHIELDS

4x4 FRAME CUT TO   (NOT
    APPROPRIATE TEE   PINE!)
    SHAPE.

TRIPLE GLAZING...+ GLAZING TAPE,
                    SETTING BLOCK

WOOD STOPS, NAILED.
WOOD PLUGS COVER
    RECESSED SCREWS.

MULLION - 2

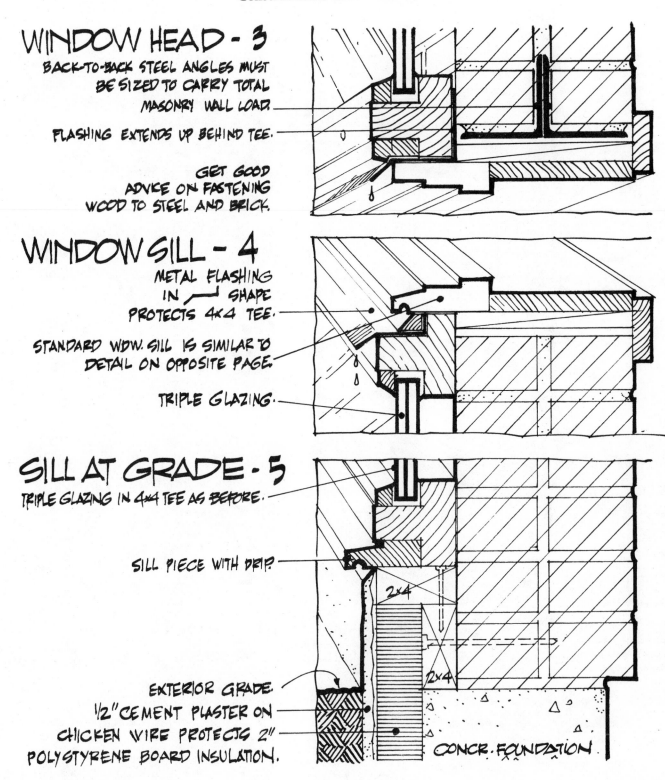

WINDOW HEAD - 3

BACK-TO-BACK STEEL ANGLES MUST BE SIZED TO CARRY TOTAL MASONRY WALL LOAD.

FLASHING EXTENDS UP BEHIND TEE.

GET GOOD ADVICE ON FASTENING WOOD TO STEEL AND BRICK.

WINDOW SILL - 4

METAL FLASHING IN ⌐ SHAPE PROTECTS 4x4 TEE.

STANDARD WDW. SILL IS SIMILAR TO DETAIL ON OPPOSITE PAGE.

TRIPLE GLAZING.

SILL AT GRADE - 5

TRIPLE GLAZING IN 4x4 TEE AS BEFORE.

SILL PIECE WITH DRIP.

2x4

2x4

EXTERIOR GRADE.

1/2" CEMENT PLASTER ON CHICKEN WIRE PROTECTS 2" POLYSTYRENE BOARD INSULATION.

CONCR. FOUNDATION

Appendix 1

Sun Path Diagrams*

Sun path diagrams are representations on a flat surface of the sun's path across the sky. They are used to easily and quickly determine the location of the sun at any time of the day and at any time of the year. Each latitude has its own sun path diagrams.

The horizon is represented as the outer circle, with you in its center. The concentric circles represent the angle of the sun above the horizon, that is, its height in the sky. The radial lines represent its angle relative to due south.

The paths of the sun on the 21st day of each month are the elliptical curves. Roman numerals label the curves for the appropriate months. For example, curve III (March) is the same as curve IX (September). The vertical curves represent the time of day. Morning is on the right (east) side of the diagrams and afternoon on the left (west).

---

* Courtesy *Architectural Graphic Standards* by C.G. Ramsey and H.R. Sleeper, John Wiley & Sons, New York, N.Y. 1972.

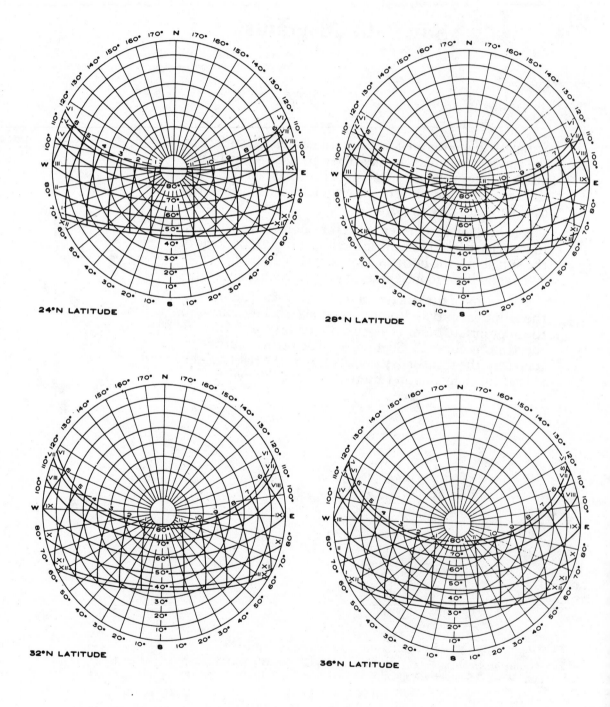

24°N LATITUDE

28° N LATITUDE

32°N LATITUDE

36°N LATITUDE

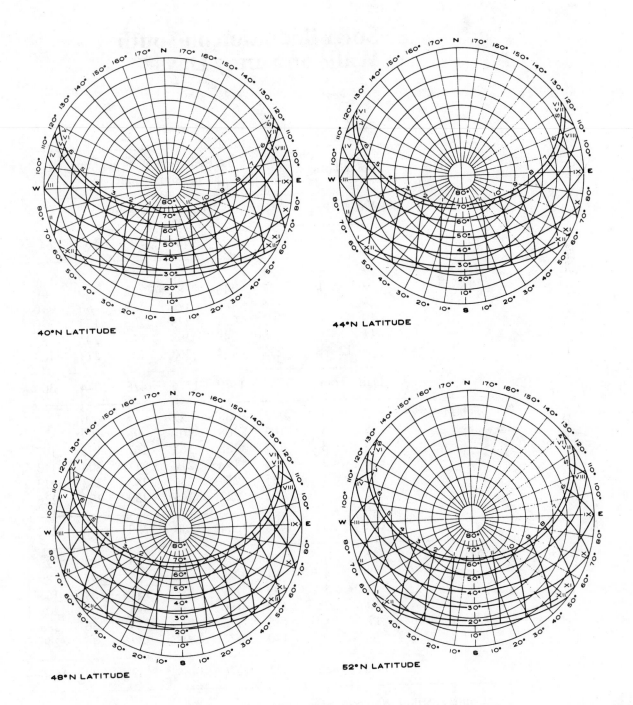

40°N LATITUDE

44°N LATITUDE

48°N LATITUDE

52°N LATITUDE

Appendix 2

Solar Radiation on South Walls on Sunny Days *

Date	Solar Time		Solar Radiation, Btus per hour per square foot					
	AM	PM	Latitude (North)					
			24	32	40	48	56	64
Jan 21	7	5	31	1	0	0	0	0
	8	4	127	115	84	22	0	0
	9	3	176	181	171	139	60	0
	10	2	207	221	223	206	153	20
	11	1	226	245	253	243	201	81
		12	232	253	263	255	217	103
	Daily Totals		1766	1779	1726	1478	1044	304
Feb 21	7	5	46	38	22	4	0	0
	8	4	102	108	107	96	69	19
	9	3	141	158	167	167	151	107
	10	2	168	193	210	217	208	173
	11	1	185	214	236	247	243	213
		12	191	222	245	259	255	226
	Daily Totals		1476	1644	1730	1720	1598	1252
March 31	7	5	27	32	35	35	32	25
	8	4	64	78	89	96	97	89
	9	3	95	119	138	152	154	153
	10	2	120	150	176	195	205	203
	11	1	135	170	200	223	236	235
		12	140	177	208	232	246	246
	Daily Totals		1022	1276	1484	1632	1700	1656

* Courtesy ASHRAE, *Handbook of Fundamentals.*

Date	Solar Time		Solar Radiation, Btus per hour per square foot					
	AM	PM	Latitude (North)					
			24	32	40	48	56	64
Apr 21	6	6	2	3	4	5	6	6
	7	5	10	10	12	21	29	37
	8	4	16	35	53	69	82	91
	9	3	41	68	93	115	133	145
	10	2	61	95	126	152	174	188
	11	1	74	112	147	177	266	216
		12	79	118	154	185	209	225
	Daily Totals		488	764	1022	1262	1458	1544
May 21	6	6	5	7	9	10	11	11
	7	5	12	13	13	13	16	28
	8	4	15	15	25	45	63	80
	9	3	16	33	60	86	109	128
	10	2	22	56	89	120	146	167
	11	1	34	72	108	141	170	193
		12	37	77	114	149	178	201
	Daily Totals		246	469	724	982	1218	1436
June 21	6	6	7	9	10	12	12	13
	7	5	13	14	14	15	15	23
	8	4	16	16	16	35	55	73
	9	3	18	19	47	74	98	119
	10	2	18	41	74	105	133	157
	11	1	19	56	92	126	156	181
		12	22	60	98	133	168	189
	Daily Totals		204	370	610	874	1126	1356

Date	Solar Time		Solar Radiation, Btus per hour per square foot					
	AM	PM	Latitude (North)					
			24	32	40	48	56	64
July 21	6	6	6	8	9	11	12	12
	7	5	13	14	14	14	15	28
	8	4	16	16	24	43	61	77
	9	3	18	31	58	83	106	124
	10	2	21	54	86	116	142	162
	11	1	32	69	104	137	165	187
	12		36	74	111	144	173	195
	Daily Totals		246	458	702	956	1186	1400
Aug 21	6	6	2	4	5	6	7	7
	7	5	11	12	12	20	28	35
	8	4	16	33	50	65	78	87
	9	3	39	65	89	110	126	138
	10	2	58	91	120	146	166	179
	11	1	71	107	140	169	191	205
	12		75	113	147	177	200	215
	Daily Totals		470	736	978	1208	1392	1522
Sept 21	7	5	26	30	32	31	28	21
	8	4	62	75	84	90	89	81
	9	3	93	114	132	143	147	141
	10	2	116	145	168	185	193	189
	11	1	131	164	192	212	223	220
	12		136	171	200	221	233	230
	Daily Totals		992	1226	1416	1596	1594	1532

Date	Solar Time		Solar Radiation, Btus per hour per square foot					
	AM	PM	Latitude (North)					
			24	32	40	48	56	64
Oct 21	7	5	42	32	16	1	0	0
	8	4	99	104	100	87	57	10
	9	3	138	153	160	154	138	90
	10	2	165	188	203	207	195	155
	11	1	182	209	229	237	230	295
		12	188	217	238	247	241	208
	Daily Totals		1442	1588	1654	1626	1480	1106
Nov 21	7	5	29	1	0	0	0	0
	8	4	129	111	81	22	0	0
	9	3	172	176	167	135	58	0
	10	2	204	217	219	201	148	21
	11	1	222	241	248	238	196	79
		12	228	249	258	250	211	100
	Daily Totals		1730	1742	1686	1442	1016	300
Dec 21	7	5	14	0	0	0	0	0
	8	4	130	107	56	0	0	0
	9	3	184	183	163	109	4	0
	10	2	217	226	221	190	103	0
	11	1	236	251	252	231	164	4
		12	243	259	263	244	182	17
	Daily Totals		1808	1794	1646	1304	722	24

Appendix 3

Maps of the Average Percentage of the Time the Sun Is Shining

JANUARY

FEBRUARY

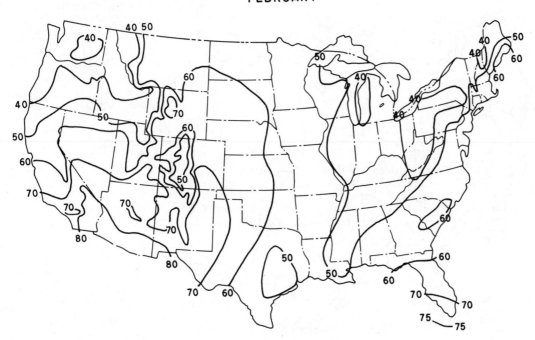

MARCH

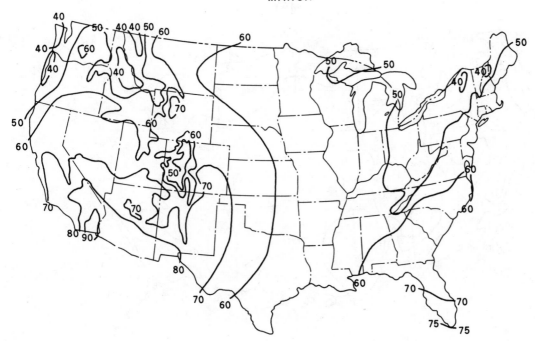

APRIL

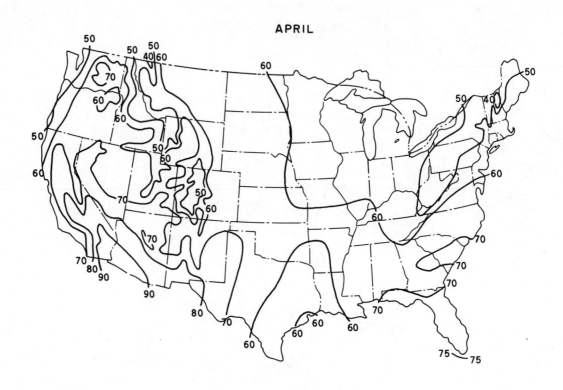

MAY

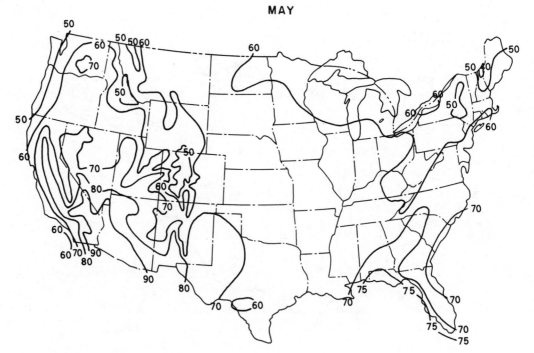

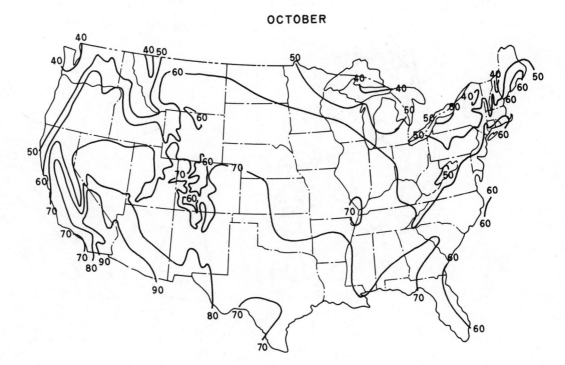

NOVEMBER

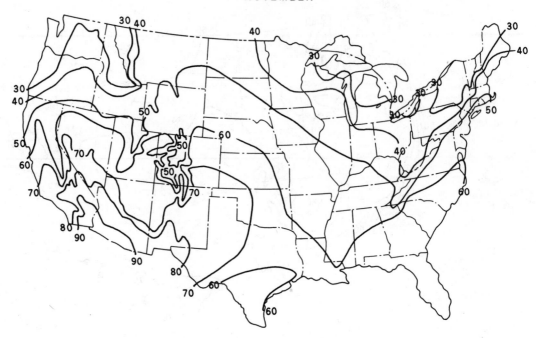

DECEMBER

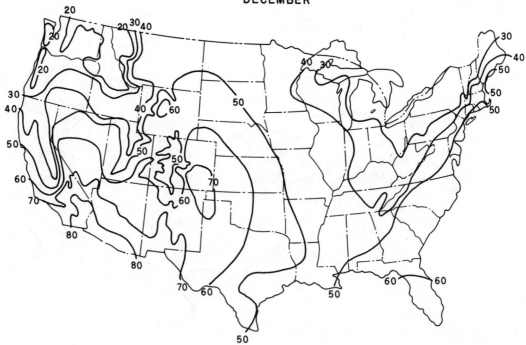

Appendix 4

U-Values for Windows with Insulating Covers

U-Values (Btus per hour per square foot per °F)

	Single Glazed Windows	Double Glazed Windows	Triple Glazed Windows
1. Window without insulation	1.15	0.55	0.35
2. With R-4 insulating cover	0.21	0.17	0.14
2a. Average U-value with R-4 cover in place 16 hr/day (¾ of the degree days)	0.45	0.27	0.20
2b. Average U-value with R-4 cover in place 12 hr/day (⅔ of the degree days)	0.52	0.29	0.21
3. With R-10 insulating cover	0.09	0.085	0.078
3a. Average U-value with R-10 cover in place 16 hr/day (¾ of the degree days)	0.36	0.20	0.15
3b. Average U-value with R-10 cover in place 12 hr/day (⅔ of the degree days)	0.44	0.24	0.17

Appendix 5

Degree Days and Design Temperatures *

*Courtesy ASHRAE, *Handbook of Fundamentals*, 1972.

State	City	Avg. Winter Temp	Design Temp	Sep	Oct	Nov	Dec	Jan	Feb	Mar	Apr	May	Yearly Total
Ala.	Birmingham	54.2	19	6	93	363	555	592	462	363	108	9	2551
	Huntsville	51.3	13	12	127	426	663	694	557	434	138	19	3070
	Mobile	59.9	26	0	22	213	357	415	300	211	42	0	1560
	Montgomery	55.4	22	0	68	330	527	543	417	316	90	0	2291
Alaska	Anchorage	23.0	−25	516	930	1284	1572	1631	1316	1293	879	592	10864
	Fairbanks	6.7	−53	642	1203	1833	2254	2359	1901	1739	1068	555	14279
	Juneau	32.1	− 7	483	725	921	1135	1237	1070	1073	810	601	9075
	Nome	13.1	−32	693	1094	1455	1820	1879	1666	1770	1314	930	14171
Ariz.	Flagstaff	35.6	0	201	558	867	1073	1169	991	911	651	437	7152
	Phoenix	58.5	31	0	22	234	415	474	328	217	75	0	1765
	Tucson	58.1	29	0	25	231	406	471	344	242	75	6	1800
	Winslow	43.0	9	6	245	711	1008	1054	770	601	291	96	4782
	Yuma	64.2	37	0	0	108	264	307	190	90	15	0	974
Ark.	Fort Smith	50.3	9	12	127	450	704	781	596	456	144	22	3292
	Little Rock	50.5	19	9	127	465	716	756	577	434	126	9	3219
	Texarkana	54.2	22	0	78	345	561	626	468	350	105	0	2533
Calif.	Bakersfield	55.4	31	0	37	282	502	546	364	267	105	19	2122
	Burbank	58.6	36	6	43	177	301	366	277	239	138	81	1646
	Eureka	49.9	32	258	329	414	499	546	470	505	438	372	4643
	Fresno	53.3	28	0	84	354	577	605	426	335	162	62	2611
	Long Beach	57.8	36	9	47	171	316	397	311	264	171	93	1803
	Los Angeles	57.4	41	42	78	180	291	372	302	288	219	158	2061
	Oakland	53.5	35	45	127	309	481	527	400	353	255	180	2870
	Sacramento	53.9	30	0	56	321	546	583	414	332	178	72	2502
	San Diego	59.5	42	21	43	135	236	298	235	214	135	90	1458
	San Francisco	55.1	42	102	118	231	388	443	336	319	279	239	3001
	Santa Maria	54.3	32	96	146	270	391	459	370	363	282	233	2967
Colo.	Alamosa	29.7	−17	279	639	1065	1420	1476	1162	1020	696	440	8529
	Colorado Springs	37.3	− 1	132	456	825	1032	1128	938	893	582	319	6423
	Denver	37.6	− 2	117	428	819	1035	1132	938	887	558	288	6283
	Grand Junction	39.3	8	30	313	786	1113	1209	907	729	387	146	5641
	Pueblo	40.4	− 5	54	326	750	986	1085	871	772	429	174	5462
Conn.	Bridgeport	39.9	4	66	307	615	986	1079	966	853	510	208	5617
	Hartford	37.3	1	117	394	714	1101	1190	1042	908	519	205	6235
	New Haven	39.0	5	87	347	648	1011	1097	991	871	543	245	5897
Del.	Wilmington	42.5	12	51	270	588	927	980	874	735	387	112	4930
D. C.	Washington	45.7	16	33	217	519	834	871	762	626	288	74	4224
Fla.	Daytona Beach	64.5	32	0	0	75	211	248	190	140	15	0	879
	Fort Myers	68.6	38	0	0	24	109	146	101	62	0	0	442
	Jacksonville	61.9	29	0	12	144	310	332	246	174	21	0	1239
	Key West	73.1	55	0	0	0	28	40	31	9	0	0	108
	Lakeland	66.7	35	0	0	57	164	195	146	99	0	0	661
	Miami	71.1	44	0	0	0	65	74	56	19	0	0	214
	Miami Beach	72.5	45	0	0	0	40	56	36	9	0	0	141
	Orlando	65.7	33	0	0	72	198	220	165	105	6	0	766
	Pensacola	60.4	29	0	19	195	353	400	277	183	36	0	1463
	Tallahassee	60.1	25	0	28	198	360	375	286	202	36	0	1485

State	City	Avg. Winter Temp	Design Temp	Sep	Oct	Nov	Dec	Jan	Feb	Mar	Apr	May	Yearly Total
	Tampa	66.4	36	0	0	60	171	202	148	102	0	0	683
	West Palm Beach	68.4	40	0	0	6	65	87	64	31	0	0	253
Ga.	Athens	51.8	17	12	115	405	632	642	529	431	141	22	2929
	Atlanta	51.7	18	18	124	417	648	636	518	428	147	25	2961
	Augusta	54.5	20	0	78	333	552	549	445	350	90	0	2397
	Columbus	54.8	23	0	87	333	543	552	434	338	96	0	2383
	Macon	56.2	23	0	71	297	502	505	403	295	63	0	2136
	Rome	49.9	16	24	161	474	701	710	577	468	177	34	3326
	Savannah	57.8	24	0	47	246	437	437	353	254	45	0	1819
Hawaii	Hilo	71.9	59	0	0	0	0	0	0	0	0	0	0
	Honolulu	74.2	60	0	0	0	0	0	0	0	0	0	0
Idaho	Boise	39.7	4	132	415	792	1017	1113	854	722	438	245	5809
	Lewiston	41.0	6	123	403	756	933	1063	815	694	426	239	5542
	Pocatello	34.8	− 8	172	493	900	1166	1324	1058	905	555	319	7033
Ill.	Chicago	37.5	− 4	81	326	753	1113	1209	1044	890	480	211	6155
	Moline	36.4	− 7	99	335	774	1181	1314	1100	918	450	189	6408
	Peoria	38.1	− 2	87	326	759	1113	1218	1025	849	426	183	6025
	Rockford	34.8	− 7	114	400	837	1221	1333	1137	961	516	236	6830
	Springfield	40.6	− 1	72	291	696	1023	1135	935	769	354	136	5429
Ind.	Evansville	45.0	6	66	220	606	896	955	767	620	237	68	4435
	Fort Wayne	37.3	0	105	378	783	1135	1178	1028	890	471	189	6205
	Indianapolis	39.6	0	90	316	723	1051	1113	949	809	432	177	5699
	South Bend	36.6	− 2	111	372	777	1125	1221	1070	933	525	239	6439
Iowa	Burlington	37.6	− 4	93	322	768	1135	1259	1042	859	426	177	6114
	Des Moines	35.5	− 7	96	363	828	1225	1370	1137	915	438	180	6588
	Dubuque	32.7	−11	156	450	906	1287	1420	1204	1026	546	260	7376
	Sioux City	34.0	−10	108	369	867	1240	1435	1198	989	483	214	6951
	Waterloo	32.6	−12	138	428	909	1296	1460	1221	1023	531	229	7320
Kans.	Dodge City	42.5	3	33	251	666	939	1051	840	719	354	124	4986
	Goodland	37.8	− 2	81	381	810	1073	1166	955	884	507	236	6141
	Topeka	41.7	3	57	270	672	980	1122	893	722	330	124	5182
	Wichita	44.2	5	33	229	618	905	1023	804	645	270	87	4620
Ky.	Covington	41.4	3	75	291	669	983	1035	893	756	390	149	5265
	Lexington	43.8	6	54	239	609	902	946	818	685	325	105	4683
	Louisville	44.0	8	54	248	609	890	930	818	682	315	105	4660
La.	Alexandria	57.5	25	0	56	273	431	471	361	260	69	0	1921
	Baton Rouge	59.8	25	0	31	216	369	409	294	208	33	0	1560
	Lake Charles	60.5	29	0	19	210	341	381	274	195	39	0	1459
	New Orleans	61.0	32	0	19	192	322	363	258	192	39	0	1385
	Shreveport	56.2	22	0	47	297	477	552	426	304	81	0	2184
Me.	Caribou	24.4	−18	336	682	1044	1535	1690	1470	1308	858	468	9767
	Portland	33.0	− 5	195	508	807	1215	1339	1182	1042	675	372	7511
Md.	Baltimore	43.7	12	48	264	585	905	936	820	679	327	90	4654
	Frederick	42.0	7	66	307	624	955	995	876	741	384	127	5087
Mass.	Boston	40.0	6	60	316	603	983	1088	972	846	513	208	5634
	Pittsfield	32.6	− 5	219	524	831	1231	1339	1196	1063	660	326	7578
	Worcester	34.7	− 3	147	450	774	1172	1271	1123	998	612	304	6969

State	City	Avg. Winter Temp	Design Temp	Sep	Oct	Nov	Dec	Jan	Feb	Mar	Apr	May	Yearly Total
Mich.	Alpena	29.7	− 5	273	580	912	1268	1404	1299	1218	777	446	8506
	Detroit	37.2	4	87	360	738	1088	1181	1058	936	522	220	6232
	Escanaba	29.6	− 7	243	539	924	1293	1445	1296	1203	777	456	8481
	Flint	33.1	− 1	159	465	843	1212	1330	1198	1066	639	319	7377
	Grand Rapids	34.9	2	135	434	804	1147	1259	1134	1011	579	279	6894
	Lansing	34.8	2	138	431	813	1163	1262	1142	1011	579	273	6909
	Marquette	30.2	− 8	240	527	936	1268	1411	1268	1187	771	468	8393
	Muskegon	36.0	4	120	400	762	1088	1209	1100	995	594	310	6696
	Sault Ste. Marie	27.7	−12	279	580	951	1367	1525	1380	1277	810	477	9048
Minn.	Duluth	23.4	−19	330	632	1131	1581	1745	1518	1355	840	490	10000
	Minneapolis	28.3	−14	189	505	1014	1454	1631	1380	1166	621	288	8382
	Rochester	28.8	−17	186	474	1005	1438	1593	1366	1150	630	301	8295
Miss.	Jackson	55.7	21	0	65	315	502	546	414	310	87	0	2239
	Meridian	55.4	20	0	81	339	518	543	417	310	81	0	2289
	Vicksburg	56.9	23	0	53	279	462	512	384	282	69	0	2041
Mo.	Columbia	42.3	2	54	251	651	967	1076	874	716	324	121	5046
	Kansas City	43.9	4	39	220	612	905	1032	818	682	294	109	4711
	St. Joseph	40.3	− 1	60	285	708	1039	1172	949	769	348	133	5484
	St. Louis	43.1	4	60	251	627	936	1026	848	704	312	121	4900
	Springfield	44.5	5	45	223	600	877	973	781	660	291	105	4900
Mont.	Billings	34.5	−10	186	487	897	1135	1296	1100	970	570	285	7049
	Glasgow	26.4	−25	270	608	1104	1466	1711	1439	1187	648	335	8996
	Great Falls	32.8	−20	258	543	921	1169	1349	1154	1063	642	384	7750
	Havre	28.1	−22	306	595	1065	1367	1584	1364	1181	657	338	8700
	Helena	31.1	−17	294	601	1002	1265	1438	1170	1042	651	381	8129
	Kalispell	31.4	− 7	321	654	1020	1240	1401	1134	1029	639	397	8191
	Miles City	31.2	−19	174	502	972	1296	1504	1252	1057	579	276	7723
	Missoula	31.5	− 7	303	651	1035	1287	1420	1120	970	621	391	8125
Neb.	Grand Island	36.0	− 6	108	381	834	1172	1314	1089	908	462	211	6530
	Lincoln	38.8	− 4	75	301	726	1066	1237	1016	834	402	171	5864
	Norfolk	34.0	−11	111	397	873	1234	1414	1179	983	498	233	6979
	North Platte	35.5	− 6	123	440	885	1166	1271	1039	930	519	248	6684
	Omaha	35.6	− 5	105	357	828	1175	1355	1126	939	465	208	6612
	Scottsbluff	35.9	− 8	138	459	876	1128	1231	1008	921	552	285	6673
Nev.	Elko	34.0	−13	225	561	924	1197	1314	1036	911	621	409	7433
	Ely	33.1	− 6	234	592	939	1184	1308	1075	977	672	456	7733
	Las Vegas	53.5	23	0	78	387	617	688	487	335	111	6	2709
	Reno	39.3	2	204	490	801	1026	1073	823	729	510	357	6332
	Winnemucca	36.7	1	210	536	876	1091	1172	916	837	573	363	6761
N. H.	Concord	33.0	−11	177	505	822	1240	1358	1184	1032	636	298	7383
N. J.	Atlantic City	43.2	14	39	251	549	880	936	848	741	420	133	4812
	Newark	42.8	11	30	248	573	921	983	876	729	381	118	4589
	Trenton	42.4	12	57	264	576	924	989	885	753	399	121	4980
N. M.	Albuquerque	45.0	14	12	229	642	868	930	703	595	288	81	4348
	Raton	38.1	− 2	126	431	825	1048	1116	904	834	543	301	6228
	Roswell	47.5	16	18	202	573	806	840	641	481	201	31	3793
	Silver City	48.0	14	6	183	525	729	791	605	518	261	87	3705

State	City	Avg. Winter Temp	Design Temp	Sep	Oct	Nov	Dec	Jan	Feb	Mar	Apr	May	Yearly Total
N. Y.	Albany	34.6	− 5	138	440	777	1194	1311	1156	992	564	239	6875
	Binghamton	36.6	− 2	141	406	732	1107	1190	1081	949	543	229	6451
	Buffalo	34.5	3	141	440	777	1156	1256	1145	1039	645	329	7062
	New York	42.8	11	30	233	540	902	986	885	760	408	118	4871
	Rochester	35.4	2	126	415	747	1125	1234	1123	1014	597	279	6748
	Schenectady	35.4	− 5	123	422	756	1159	1283	1131	970	543	211	6650
	Syracuse	35.2	− 2	132	415	744	1153	1271	1140	1004	570	248	6756
N. C.	Asheville	46.7	13	48	245	555	775	784	683	592	273	87	4042
	Charlotte	50.4	18	6	124	438	691	691	582	481	156	22	3191
	Greensboro	47.5	14	33	192	513	778	784	672	552	234	47	3805
	Raleigh	49.4	16	21	164	450	716	725	616	487	180	34	3393
	Wilmington	54.6	23	0	74	291	521	546	462	357	96	0	2347
	Winston-Salem	48.4	14	21	171	483	747	753	652	524	207	37	3595
N. D.	Bismarck	26.6	−24	222	577	1083	1463	1708	1442	1203	645	329	8851
	Devils Lake	22.4	−23	273	642	1191	1634	1872	1579	1345	753	381	9901
	Fargo	24.8	−22	219	574	1107	1569	1789	1520	1262	690	332	9226
	Williston	25.2	−21	261	601	1122	1513	1758	1473	1262	681	357	9243
Ohio	Akron-Canton	38.1	1	96	381	726	1070	1138	1016	871	489	202	6037
	Cincinnati	45.1	8	39	208	558	862	915	790	642	294	96	4410
	Cleveland	37.2	2	105	384	738	1088	1159	1047	918	552	260	6351
	Columbus	39.7	2	84	347	714	1039	1088	949	809	426	171	5660
	Dayton	39.8	0	78	310	696	1045	1097	955	809	429	167	5622
	Mansfield	36.9	1	114	397	768	1110	1169	1042	924	543	245	6403
	Toledo	36.4	1	117	406	792	1138	1200	1056	924	543	242	6494
	Youngstown	36.8	1	120	412	771	1104	1169	1047	921	540	248	6417
Okla.	Oklahoma City	48.3	11	15	164	498	766	868	664	527	189	34	3725
	Tulsa	47.7	12	18	158	522	787	893	683	539	213	47	3860
Ore.	Astoria	45.6	27	210	375	561	679	753	622	636	480	363	5186
	Eugene	45.6	22	129	366	585	719	803	627	589	426	279	4726
	Medford	43.2	21	78	372	678	871	918	697	642	432	242	5008
	Pendleton	42.6	3	111	350	711	884	1017	773	617	396	205	5127
	Portland	45.6	21	114	335	597	735	825	644	586	396	245	4635
	Roseburg	46.3	25	105	329	567	713	766	608	570	405	267	4491
	Salem	45.4	21	111	338	594	729	822	647	611	417	273	4754
Pa.	Allentown	38.9	3	90	353	693	1045	1116	1002	849	471	167	5810
	Erie	36.8	7	102	391	714	1063	1169	1081	973	585	288	6451
	Harrisburg	41.2	9	63	298	648	992	1045	907	766	396	124	5251
	Philadelphia	41.8	11	60	297	620	965	1016	889	747	392	118	5144
	Pittsburgh	38.4	5	105	375	726	1063	1119	1002	874	480	195	5987
	Reading	42.4	6	54	257	597	939	1001	885	735	372	105	4945
	Scranton	37.2	2	132	434	762	1104	1156	1028	893	498	195	6254
	Williamsport	38.5	1	111	375	717	1073	1122	1002	856	468	177	5934
R. I.	Providence	38.8	6	96	372	660	1023	1110	988	868	534	236	5954
S. C.	Charleston	57.9	26	0	34	210	425	443	367	273	42	0	1794
	Columbia	54.0	20	0	84	345	577	570	470	357	81	0	2484
	Florence	54.5	21	0	78	315	552	552	459	347	84	0	2387
	Greenville-Spartanburg	51.6	18	6	121	399	651	660	546	446	132	19	2980
S. D.	Huron	28.8	−16	165	508	1014	1432	1628	1355	1125	600	288	8223

State	City	Avg. Winter Temp	Design Temp	Sep	Oct	Nov	Dec	Jan	Feb	Mar	Apr	May	Yearly Total
	Rapid City	33.4	− 9	165	481	897	1172	1333	1145	1051	615	326	7345
	Sioux Falls	30.6	−14	168	462	972	1361	1544	1285	1082	573	270	7839
Tenn.	Bristol	46.2	11	51	236	573	828	828	700	598	261	68	4143
	Chattanooga	50.3	15	18	143	468	698	722	577	453	150	25	3254
	Knoxville	49.2	13	30	171	489	725	732	613	493	198	43	3494
	Memphis	50.5	17	18	130	447	698	729	585	456	147	22	3232
	Nashville	48.9	12	30	158	495	732	778	644	512	189	40	3578
Tex.	Abilene	53.9	17	0	99	366	586	642	470	347	114	0	2624
	Amarillo	47.0	8	18	205	570	797	877	664	546	252	56	3985
	Austin	59.1	25	0	31	225	388	468	325	223	51	0	1711
	Corpus Christi	64.6	32	0	0	120	220	291	174	109	0	0	914
	Dallas	55.3	19	0	62	321	524	601	440	319	90	6	2363
	El Paso	52.9	21	0	84	414	648	685	445	319	105	0	2700
	Galveston	62.2	32	0	6	147	276	360	263	189	33	0	1274
	Houston	61.0	28	0	6	183	307	384	288	192	36	0	1396
	Laredo	66.0	32	0	0	105	217	267	134	74	0	0	797
	Lubbock	48.8	11	18	174	513	744	800	613	484	201	31	3578
	Port Arthur	60.5	29	0	22	207	329	384	274	192	39	0	1447
	San Antonio	60.1	25	0	31	204	363	428	286	195	39	0	1546
	Waco	57.2	21	0	43	270	456	536	389	270	66	0	2030
	Wichita Falls	53.0	15	0	99	381	632	698	518	378	120	6	2832
Utah	Milford	36.5	− 1	99	443	867	1141	1252	988	822	519	279	6497
	Salt Lake City	38.4	5	81	419	849	1082	1172	910	763	459	233	6052
Vt.	Burlington	29.4	−12	207	539	891	1349	1513	1333	1187	714	353	8269
Va.	Lynchburg	46.0	15	51	223	540	822	849	731	605	267	78	4166
	Norfolk	49.2	20	0	136	408	698	738	655	533	216	37	3421
	Richmond	47.3	14	36	214	495	784	815	703	546	219	53	3865
	Roanoke	46.1	15	51	229	549	825	834	722	614	261	65	4150
Wash.	Olympia	44.2	21	198	422	636	753	834	675	645	450	307	5236
	Seattle	46.9	28	129	329	543	657	738	599	577	396	242	4424
	Spokane	36.5	− 2	168	493	879	1082	1231	980	834	531	288	6655
	Walla Walla	43.8	12	87	310	681	843	986	745	589	342	177	4805
	Yakima	39.1	6	144	450	828	1039	1163	868	713	435	220	5941
W. Va.	Charleston	44.8	9	63	254	591	865	880	770	648	300	96	4476
	Elkins	40.1	1	135	400	729	992	1008	896	791	444	198	5675
	Huntington	45.0	10	63	257	585	856	880	764	636	294	99	4446
	Parkersburg	43.5	8	60	264	606	905	942	826	691	339	115	4754
Wisc.	Green Bay	30.3	−12	174	484	924	1333	1494	1313	1141	654	335	8029
	La Crosse	31.5	−12	153	437	924	1339	1504	1277	1070	540	245	7589
	Madison	30.9	− 9	174	474	930	1330	1473	1274	1113	618	310	7863
	Milwaukee	32.6	− 6	174	471	876	1252	1376	1193	1054	642	372	7635
Wyo.	Casper	33.4	−11	192	524	942	1169	1290	1084	1020	657	381	7410
	Cheyenne	34.2	− 6	219	543	909	1085	1212	1042	1026	702	428	7381
	Lander	31.4	−16	204	555	1020	1299	1417	1145	1017	654	381	7870
	Sheridan	32.5	−12	219	539	948	1200	1355	1154	1051	642	366	7680

Appendix 6

Selecting the Right
Glazing Material *

The most important things to consider in choosing a glazing material are appearance, durability, performance, and cost. Since the glazing is visible, whether it is clear or cloudy, shiny or dull, or flat or bowed, it dramatically affects the appearance of the system. Durability is critical since the glazing provides the outer barrier to water, cold air, ultraviolet radiation, and weather. High transmittance of light and low transmittance of heat affect the efficiency of the system. The glazing should be inexpensive and easy to handle. The table summarizes important properties of various glazings.

Glass

Glass is usually a more expensive choice but a very popular one. Although common glass is less expensive, tempered glass is stronger and safer. "Water-white" glass (fully tempered) has a very low iron-oxide content (0.01 percent) and thus the highest transmittance (0.91). Tempered float glass is less expensive but has a high iron-oxide content and a transmittance of 0.91. However, use it only vertically and when safety is a small factor.

Glass is rigid, it looks good, it's durable, and it resists weathering and chemical and light deterioration. Unfortunately, it is heavy and difficult to handle. It also breaks.

---

* Based on an article by Peter Temple and Joe Kohler entitled "Glazing Choices" in *Solar Age* magazine, April 1979, Harrisville, NH 03450.

Glass prices vary significantly depending on how much you buy and where you buy it. Tempered low-iron glass ("water-white") usually has the same retail price as float glass, roughly $2 to $2.50 per square foot. But you can buy water-white glass for as little as $1.00 per square foot if you shop hard enough.

Fiberglass-reinforced Polyester

Fiberglass-reinforced polyester (FRP) glazing materials appear cloudy, but their solar transmittance (0.84–0.90) is only slightly less than low-iron glass. Kalwall's Sun-lite™ and Vistron's Filon™ are two commercially available FRP glazings.

FRPs are available in 4- and 5-foot-wide rolls in thicknesses of 0.025, 0.040 and 0.060 inches. It is a popular material since it is easy to cut, drill, and install. Some people do not like its appearance; it does not lie flat and often looks wrinkled. Kalwall has double-glazed panels onto which the FRP is stretched taut over an aluminum frame. The panels are less wrinkled but are not entirely smooth.

FRPs degrade somewhat at high temperatures. Kalwall notes that their Sun-lite loses 1%, 3%, and 11% of its transmittance when exposed to temperatures of 150°F, 200°F, and 300°F, respectively, for 300 hours. Most passive applications do not reach 200°. Tilted convective loop collectors are the main exception.

Filon is an acrylic-fortified polyester, reinforced with fiberglass. A thin layer of Tedlar™ polyvinylfluoride provides protection from ultraviolet degradation and weathering. Filon is available in flat or corrugated sheets. The corrugations reduce the wavy appearance problem. Filon, like Sun-lite, may require venting in higher temperature applications to protect it from thermal degradation.

Films

Plastic films are very transparent and are relatively inexpensive. Two of the best materials are Dupont's Teflon™ and Tedlar™. Teflon stands up well in high temperatures except that it expands and sags. It is difficult to handle, bowing between supports and sticking to surfaces like Saran Wrap™.

Tedlar is also difficult to handle and install. Dupont recommends that it be used only at low temperatures. Direct exposure to ultraviolet radiation causes embrittlement, and this effect is tremendously accelerated at higher temperatures. Used at low temperatures, Tedlar has an expected lifetime of 4 to 5 years until embrittlement. If there are any hotspots (e.g., near a hot metal support), these places will embrittle earlier. A new version of Tedlar, 400xRB160SE, has recently been developed, and it is expected that this product will be less susceptible to UV degradation.

A new product by 3M Company, Flexiguard™ 7410, may avoid one of the problems of most films. The manufacturer claims that it does not sag at high temperatures. It remains rigid, but not brittle, at temperatures from −30°F to 3000°F.

A common disadvantage of thin plastic films is their transparency to long-wave radiation (heat). The resulting higher heat loss reduces efficiency. Glass has a transmittance of heat of less than 1%, but the transmittance for films ranges from 17% for 5 mil polyester to 30% for 4 mil Teflon REP and 57% for 1 mil Tedlar. Long-wave transmittance data for Flexiguard 7410 is not presently available.

Rigid Plastics

Rigid plastic glazings are strong, easy to handle, and generally attractive. Most of them are either acrylics or polycarbonates. Acrylics are slightly more transparent than tempered water-white glass and resist ultraviolet light and weathering. They are usually clear and are as attractive as glass if they are not scratched. They tend to soften and bow at higher temperatures, but this is not a concern for most passive applications.

Polycarbonates are stronger than acrylics, but they have a lower transmittance and suffer from ultraviolet degradation. Like acrylics, polycarbonates have a high coefficient of thermal expansion and bow inward when the passive system gets too hot.

Insulating Panels

Some glazing materials are manufactured as "insulating" panels, which form a rigid sandwich: an air space between two glazing layers. Their higher initial cost may be offset by the lower installation cost compared with two individually installed layers.

Comparing Glazings

	Thickness (in.)	Cost ($/ft.²)	Transmittance	Weight/Area (lb/ft.²)	Thermal Expansion (°F⁻¹ x10⁻⁵)	Ease in Handling	Strength	Sheet Size (ft.)	Remarks
Water white glass "Solatex" (ASG)	0.125	0.99	0.90	1.60	0.47	Poor	Good (tempered)	2, 3, or 4x8	Very durable—no degradation
Float glass	0.125	2.35	0.84	1.60	0.47	Poor	Good (tempered)	4x8	Very durable—no degradation
Window glass (ASG SS Lustra-glass)	0.090	1.80	0.91	1.20	0.47	Poor	Poor (non-tempered)	4x7	Fragile
Sunlite Premium II (Kalwall)	0.040	0.60	0.88	0.29	2.00	Excellent	Very good	4 or 5 width rolls	Maximum temperature 300°F
Filon w/Tedlar (Vistron Corp.)	--	1.00	0.86	0.25	2.30	Very good	Very good	4.25x16	Maximum temperature 300°F
Flexiguard 7410 (3M)	7 mil	0.38	0.89	0.053	--	Fair	Good	4x150 roll	Maximum temperature 275°F
Tedlar (Dupont)	4 mil	0.05	0.95	0.029	2.80	Fair	Good, some embrittlement	up to 5.33 width roll (64 in.)	4-5 yr. lifetime at 150°F
Teflon FEP 100A (Dupont)	1 mil	0.58	0.96	0.02	5.85	Poor	Fair, not for exterior glazing	4.83 width roll (58 in.)	Maximum temperature 300°F
Swedcast 300 Acrylic (Swedlow Inc.)	0.125	0.81	0.93	0.77	4	Excellent	Very good	9 wide	Maximum service temperature 200°F
Lucite Acrylic (Dupont)	0.125	1.14	0.92	0.73	4	Very good	Very good	4x8	Maximum temperature 200°F
Tuffak-Twinwall (Rhom & Haas)	--	1.25 (2 layers)	Equiv. to 0.89 for 1 layer	0.25	3.3	Very good	High impact strength fatigue cracking	4x8	5% reduction in transmittance over 5 years
Acrylite SDP (Cyro)	--	2.15 (2 layers)	Equiv. to 0.93 for 1 layer	1.00	4	Very good	Good	6x8	Maximum temperature 230°F
Sun-lite Insulated Panels (Kalwall)	--	2.50 (2 layers)	Equiv. to 0.88 for 1 layer	0.7	--	Good	Good	4x8 4x10 4x12 4x14	Maximum temperature 300°F
Solar Glass Panels (ASG)	--	2.99 (2 layers)	Equiv. to 0.90 for 1 layer	4.5	0.47	Poor	Good	3 or 4x6 3 or 4x8	Very durable

---

* Courtesy *Solar Age* magazine, Harrisville, NH.

Tuffak-Twinwall™ is a sandwich of polycarbonate material. Although it is relatively inexpensive, it has the same serious disadvantages of any polycarbonate: ultraviolet degradation, low transmittance, and a large coefficient of expansion. Likewise, the Cyro-Acrylic SDP™ panels have the disadvantages associated generally with acrylics: a low melting point and a large coefficient of thermal expansion.

ASG sells double solar glass panels using either their Solatex™ or Sunadex™ glazing. These panels are designed specifically for solar applications. The two layers of glass are hermetically sealed.

Directory of
Passive Solar Resources

Introduction

The purpose of this section is to help you obtain products
and further information. Be informed before beginning any
solar endeavor—from designing and building a house to
organizing an educational course or program. Rarely is
it necessary to start from "scratch" in researching a solar
project because a lot of free or inexpensive information
already exists. We want to help you find what you need
easily. Many direct sources are included, but to list all
available solar sources would require many pages. Instead,
we have provided major solar information centers and
organizations. Our hope is that you will find this resource
complete, yet fairly compact and easy to use.

Information Sources

National Solar Heating and Cooling Information Center (NSHCIC)

Established by the federal government, this information
center provides many free lists of solar designers, builders,
installers, organizations, construction plans, and
bibliographies including audiovisuals. This is an excellent
first step for access to information on a variety of solar
subjects. Write to

Solar Heating
P.O. Box 1607
Rockville, MD 20850

or call free:
(800) 523-2929
(800) 462-4983 in Pennsylvania
(800) 523-4700 in Alaska and Hawaii

State Energy Offices *

Each state now has an energy office that is in many cases one of the best sources for local information about educational materials, organizations, technical assistance, and reliable professional solar services. The following is a list of state energy offices.

ALABAMA
Alabama Solar Energy Center
University of Alabama/
Huntsville
P.O. Box 1247
Huntsville, AL 35807
(205) 895-6361
(800) 572-7226
Solar products and services catalog. TVA passive solar home plans; construction details for individual problems; answers design questions on an individual basis; instructional materials for teachers and do-it-yourselfers.
Alabama Energy Extension Service
Auburn University
Auburn, AL 36830
(205) 826-4718
ALASKA
Department of Commerce and Economic Development Division of Energy and Power Development
338 Denali Street
Anchorage, AK 99501
(907) 276-0508

Outlets across the state—will direct inquirers to local sources of information.
Alaska Energy Extension Service
7th Floor
MacKay Bldg.
338 Denali Street
Anchorage, AK 99501
(907) 274-8655

ARIZONA
Arizona Solar Energy Commission
1700 West Washington Street
Phoenix, AZ 85007
(602) 255-3682
(800) 325-5499
Passive solar home designs. Conferences and workshops. State solar directory of equipment and services.

ARKANSAS
Arkansas Department of Energy
3000 Kavanaugh Boulevard
Little Rock, AR 72205
(501) 371-1370
(800) 325-5499

Design and construction manuals for a greenhouse, an air space-heating system, and a solar water heating system available by late fall 1980.
CALIFORNIA
Energy Information
California Energy Commission
1111 Howe Avenue
Sacramento, CA 95825
(916) 920-6430
(800) 852-7516
Excellent solar business directory (available from Business and Transportation Agency. Solar Business Office, 921 10th Street, Sacramento, Calif. 95814); a publications catalog; passive solar design manual.
California Department of Consumer Affairs
Solar Unit
1020 N Street
Sacramento, CA 95814
(916) 322-5756
(800) 952-5567
Mediating group between consumers and industry.

---

* Courtesy of *Solar Age*, SolarVision, Inc., Harrisville, New Hampshire.

California Energy Extension Service
Office of Appropriate Technology
1530 Tenth Street
Sacramento, CA 95814
(916) 445-1803
(916) 322-8901
Excellent publications on solar and renewable energy written for consumers.

COLORADO
Office of Energy Conservation
1600 Downing Street
Second Floor
Denver, CO 80218
(303) 839-2507
Solar business and professional organization directory.
Energy Environment Information Center
Denver Public Library
1357 Broadway
Denver, CO 80203
(303) 837-5994
Building plans, construction details, design aids, equipment catalogs, books, etc. Most have to be used at the library.

CONNECTICUT
Solar Office of Policy and Management
80 Washington Street
Hartford, CT 06115
(203) 566-5803

DELAWARE
Governor's Energy Office
P.O. Box 1401
114 West Water Street
Dover, DE 19901
(302) 678-5644
(800) 282-8616

FLORIDA
Florida Solar Energy Center
300 State Road 401

Cape Canaveral, FL 32920
(305) 783-0300
Industry directory; manual for solar installers; teachers' workshops; developing curricula for grades K-12; quarterly newsletter; consumer's guide to solar water heating.
Governor's Energy Office
301 Bryant Building
Tallahassee, FL 32304
(904) 488-2475

GEORGIA
Georgia Office of Energy Resources
Room 615
270 Washington Street, S.W.
Atlanta, GA 30334
(404) 656-5176

HAWAII
Hawaii State Energy Office
Department of Planning and Economic Development
P.O. Box 2359
Honolulu, HI 96804
(808) 548-4150
State Energy Hotlines: Oahu 548-4080; other islands call Enterprise Operator 8016: Solar/Wind Handbook for Hawaii; directory of solar water heating companies; monthly newsletter; will direct calls to decentralized energy extension services; booklet on building and installing a solar domestic water heating system.

IDAHO
Office of Energy
Western SUN Idaho
State House
Boise, ID 83720
(208) 334-3800
Handsome booklet, "Energy Alternatives for Idaho."

ILLINOIS
Institute of Natural Resources
Solar Section
325 West Adams
Springfield, IL 62706
(217) 785-2800
Building plans; a staff architect to give technical design assistance; F-CHART design program; solar business directory.

INDIANA
Indiana Department of Commerce
Solar Office
440 North Meridian
Indianapolis, IN 46204
(317) 232-8940
Solar business directory.

IOWA
Iowa Energy Policy Council
Iowa Solar Office
6th Floor
Lucas Bldg.
State Capitol Complex
Des Moines, IA 50319
(515) 281-4420
(800) 532-1114
Compiling building plans; catalogs of solar components; newsletter.

KANSAS
Kansas Energy Office
214 West Sixth
Topeka, KS 66603
(913) 296-2496
(800) 432-3537

KENTUCKY
Bureau of Energy Management
Kentucky Department of Energy
Iron Works Pike
Lexington, KY 40578
(606) 252-5535
(800) 432-9014

At least two library shelves, ceiling to floor, of catalogs and directories.

LOUISIANA
Research and Development Division Department of Natural Resources
P.O. Box 44156
Baton Rouge, LA 70804
(504) 342-4594
Solar business, professional, and organization directory. Construction details. Developing design manual for design types, costs, and evaluation procedures for solar energy systems.

MAINE
Maine Office of Energy Resources
State House
Station Number 53
Augusta, ME 04330
(207) 289-3811
(800) 452-4648
Plans for thermosiphoning collectors, breadbox/wood water heater. Professional solar services directory.

MARYLAND
Maryland Energy Office
Room 1302
301 West Preston Street
Baltimore, MD 21201
(301) 383-6810
(800) 492-5903

MASSACHUSETTS
Executive Office of Energy Resources
Renewable Energy Division
73 Tremont Street
Room 849
Boston, MA 02108
(617) 727-7297
(800) 922-8265

Building plans for vertical wall collector and solar greenhouse.

MICHIGAN
Michigan Department of Commerce Energy Administration
P.O. Box 30228
Lansing, MI 48909
(517) 373-6430
(800) 292-4704

MINNESOTA
Minnesota Energy Agency
980 American Center Bldg.
150 East Kellog
St. Paul, MN 55101
(612) 296-5120
(800) 652-9747
Building plans and construction details. Buyer's guide.

MISSISSIPPI
Governor's Office of Energy & Transportation
510 George Street
Jackson, MS 39205
Office in transition. Little available at the present time.

MISSOURI
Missouri Division of Energy
Missouri Department of Natural Resources
P.O. Box 1309 (for general information)
P.O. Box 176 (for solar information)
Jefferson City, MO 65102
(314) 751-4000
(800) 392-0717
Building plans for passive water heaters, passive homes, greenhouses, window air-heaters, list of manufacturers and distributors, will evaluate sets of plans sent to the office.

MONTANA
Energy Division
Department of Natural Resources and Conservation
32 South Ewing
Helena, MT 59601
(406) 449-4624
Building plans, residential, commercial, institutional, greenhouses, swimming pool heaters, and more; maintains a geothermal division; contact Jeff Berkeley.

NEBRASKA
Nebraska Solar Office
W-191
Nebraska Hall
University of Nebraska
Lincoln, NE 68588
(402) 472-3414
Preparing workbooks.

NEVADA
Nevada Department of Energy
400 West King Street
Carson City, NV 89710
(702) 885-5157
(800) 992-0900
Directory of solar businesses and professionals.

NEW HAMPSHIRE
Solar
Governor's Council on Energy
2½ Beacon Street
Concord, NH 03301
(603) 271-2711
(800) 852-3311
Solar business directory. Solar greenhouse plans.

NEW JERSEY
Office of Alternative Technology
New Jersey State Energy Office
101 Commerce Street
Newark, NJ 07102

(201) 648-6293
(800) 492-4242
Newsletter; equipment supplier's list.

NEW MEXICO
Solar
Energy and Minerals
Department
P.O. Box 2770
Santa Fe, NM 87501
(505) 827-2472
(800) 432-6782/6783
Solar business directory; newsletter.
New Mexico Solar Energy
Extension Service
Box 00
Santa Fe, NM 87501
(505) 827-5621

NEW YORK
New York State Energy Office
Solar Program
2 Rockefeller Plaza
Albany, NY 12223
(518) 474-7016
(800) 342-3722
Energy conservation workbooks; training manual for energy-conscious construction.
New York State Research and
Development Authority
Agency Building 2
Rockefeller Plaza
Albany, NY 12223
(518) 465-6251
Seventeen passive solar building designs.

NORTH CAROLINA
Information Section
Energy Division
North Carolina Department of
Commerce
430 North Salisbury Street
Raleigh, NC 27611

(919) 733-2230
Newsletter.

NORTH DAKOTA
FAC Office
Energy Management and
Conservation Programs
1533 North 12th Street
Bismarck, ND 58501
(701) 224-2250

OHIO
Solar Information
Ohio Department of Energy
30 East Broad Street
34th Floor
Columbus, OH 43215
(614) 466-1805
(800) 282-9234
Passive design aids. Plans for a solar still.

OKLAHOMA
Oklahoma Department of
Energy
4400 North Lincoln Blvd.
Suite 251
Oklahoma City, OK 73105
(405) 521-3941

OREGON
Oregon Department of Energy
Room 111
Labor and Industries Bldg.
Salem, OR 97310
(503) 378-4040
(800) 452-7813
Attractive publication, "The Oregon Sunbook."
Oregon Extension Services
(There are seven different extension service offices in the state—to get the address and phone number of the one nearest you, call the above state energy office number.)
Plan Service
Department of Agricultural
Engineering

Oregon State University
Gilmore Hall
Corvallis, OR 97331
Developing plans for an energy-efficient house and a solar house; Plans for a solar greenhouse and the USDA solar home.

PENNSYLVANIA
Governor's Energy Council
1625 North Front Street
Harrisburg, PA 17101
(717) 783-8610
(800) 882-8400

RHODE ISLAND
Governor's Energy Office
80 Dean Street
Providence, RI 02903
(401) 277-3773/3774
(people in state may call collect)
Greenhouse plans; manufacturers literature file.

SOUTH CAROLINA
Division of Energy Resources
1122 Lady Street
11th Floor
Columbia, SC 29201
(803) 758-8110
(800) 922-5310
Governor's Office of Economic
Opportunity
State Economic Opportunity
Office of the State of South
Carolina
1712 Hampton Street
Columbia, SC 29201
(803) 758-3191
Audiovisuals on how to build space-heating and domestic hot water systems.

SOUTH DAKOTA
State Solar Office
Suite 102
Capitol Lake Plaza Bldg.
Pierre, SD 57501

(605) 773-3603
(800) 592-1865
Newsletter; distributors list.
TENNESSEE
Tennessee Energy Authority
Suite 707
Capitol Building
226 Capitol Blvd.
Nashville, TN 37219
(800) 342-1340
Conservation guide for grades 7-12; refers inquiries to TVA, Oak Ridge Technical Information Center.
TEXAS
Information Service
Texas Energy and National Resources Advisory Council
411 West 13th Street
Room 800
Austin, TX 78701
(512) 475-5588
UTAH
Utah Energy Office
231 East 400 South
Suite 101
Salt Lake City, UT 84111
(801) 581-5424
(800) 662-3633
Building plans for solar food dehydrator. Film on how to build a solar greenhouse.

VERMONT
Solar
State Energy Office
4 East State Street
Montpelier, VT 05602
(802) 828-2393
(800) 642-3281
Newsletter; solar products guide; greenhouse plans.

VIRGINIA
Virginia Energy Information and Services Center
310 Turner Road
Richmond, VA 23225
(804) 745-3245
(800) 552-3831

WASHINGTON
State Energy Office
400 East Union Street
First Floor
Olympia, WA 98504
(206) 754-1350
Energy Extension Service
Washington State University
Room 312
Smith Tower
Seattle, WA 98104
(206) 344-7984
Newsletter; offers classes on a variety of solar and renewable energy subjects.

WEST VIRGINIA
West Virginia Fuel and Energy Office
1262½ Greenbriar Road
Charleston, WV 25311
(304) 348-8860
(800) 642-9012
Solar business directory; some building plans.

WISCONSIN
Wisconsin Division of State Energy
101 South Webster
8th Floor
Madison, WI 53702
(608) 266-8234

WYOMING
Wyoming Energy Extension Service
P.O. Box 3965
University Station
Laramie, WY 82070
(307) 766-3362
(800) 442-6783
Plans for a solar greenhouse geared to Wyoming climate; plans for a window-box collector, solar food dehydrator, attached solar greenhouse for mobile homes.

Audio Visuals

Complete listings of solar films, slides, and video cassettes are available from the following sources:

Department of Energy Film Library
Technical Information Center
P.O. Box 62
Oak Ridge, TN 37830

Solar Age Magazine
August 1980, $3.50
c/o SolarVision, Inc.
100 Church Hill
Harrisville, NH 03450

NSHCIC (National Solar Heating and Cooling Information Center)
P.O. Box 1607
Rockville, MD 20850
or call free (800) 523-2929
(800) 462-4983 in Pennsylvania
(800) 523-4700 in Alaska and Hawaii

SUNDRIES
c/o Solar Lobby
1001 Connecticut Avenue, N.W.
Suite 510
Washington, D.C. 20036
    Solar Energy Education Bibliography
    $3.75 plus handling: 15% in U.S.A., 20% elsewhere

Periodicals/Publishers

Solar Age
SolarVision, Inc.
100 Church Hill
Harrisville, NH 03450
Monthly, $20.00/yr.
Articles on developments in all aspects of solar energy applications with an emphasis on solar heating and cooling. Written primarily for building professionals (architects, engineers, builders), business people and educators. Bruce Anderson is the executive editor of this magazine.

Alternative Sources of Energy
Alternative Sources of Energy, Inc.
Route 2, Box 90A
Milaca, MN 56353
Quarterly, $10.00/yr.
Articles, columns, and features on many aspects of energy alternatives; serves as a clearing house for exchange of ideas and technologies.

Home Energy Digest and Wood Burning Quarterly
Home Energy Digest
Division of Investment Rarities, Inc.
8009 34th Avenue
South Minneapolis, MN 55420
Quarterly, $7.94/yr.
Feature articles and columns on home heating and energy conservation; includes solar and other energy alternatives.

The Mother Earth News
The Mother Earth News, Inc.
105 Stoney Mountain Road
Hendersonville, NC 28739
Bi-monthly, $12.00/yr.
Down-to-earth descriptions of people's experiences with alternative lifestyles, ecology, and energy; source for what is happening in energy at the grass-roots level.

Solar Energy
Pergamon Press, Inc.
Maxwell House
Fairview Park
Elmsford, NY 10523
Monthly, $121.00/yr.
(Included with membership in International Solar Energy Society — $40.00.) Scientific and engineering papers on all aspects of solar energy and technology, theory, and applications.

Solar Energy Intelligence Report
Business Publishers, Inc.
P.O. Box 1067
Silver Spring, MD 20910
Weekly, $90.00/yr.
The Washington beat in solar energy; also new developments, markets, meetings.

Solar Engineering
Solar Engineering Publishers, Inc.
8435 N. Stemmons Freeway
Suite 880
Dallas, TX 75247
Monthly, $15.00/yr.

Short descriptions of activities and developments in the field of solar energy, particularly in private industry.

Solar Heating and Cooling
Gordon Publications
P.O. Box 2126-R

Morristown, NJ 07960
Monthly, $8.00/yr.
Short articles on solar heating and cooling issues, developments, and equipment; oriented to builders, developers, and manufacturers.

The following publishers and/or distributors can provide brochures and listings of their solar and alternative energy publications. *

American Section of the International Solar Energy Society, Inc. (AS/ISES)
C/O Advanced Institute for Research Technology
Highway 190 West
Killeen, Texas 76541

Ballinger Publishing Co.
17 Dunster St.
Harvard Square
Cambridge, Mass. 02138

Brick House Publishing, Inc.
34 Essex Street
Andover, Mass. 01810

Catbird Seat Books
1231 S.W. Washington
Portland, Ore. 97205

Center for Science in the Public Interest (CSPI) Energy Project
1757 South St.
Washington, D.C. 20009

Cheshire Books
121 Stanford Ave.
Menlo Park, Calif. 94025

Citizens' Energy Project
1413 K St., N.W.
Washington, D.C. 20005

Doubleday and Co., Inc.
501 Franklin Ave.
Garden City, N.Y. 11530

Ecotope Group
2332 East Madison
Seattle, Wash. 98112

Enertech Corporation
Box 420
Norwich, Vt. 05055

Environmental Action Reprint Service (EARS)
Box 545
La Veta, Colo. 81055

Friends of the Earth, Inc.
124 Spear St.
San Francisco, Calif. 94105

Garden Way Publishing
Charlotte, Vt. 05445

Institute for Local Self-Reliance
1717 18th St., N.W.
Washington, D.C. 20009

J.G. Press, Inc.
Box 351
Emmaus, Penn. 18049

John Muir Press
P.O. Box 613
Santa Fe, N.M. 87501

Lane Publishing Co.
Willow and Middlefield Rds.
Menlo Park, Calif. 94025

Lexington Books
D.C. Heath and Company
125 Spring St.
Lexington, Mass. 02173

Little, Brown, and Co.
34 Beacon St.
Boston, Mass. 02114

Maine Audubon Society
Gilsland Farm
118 Old Route One
Falmouth, Maine 04105

McGraw-Hill Book Co.
1221 Avenue of the Americas
New York, N.Y. 10020

Modern Energy and Technology Alternatives (META) Publications
P.O. Box 128
Marblemount, Wash. 98267

National Center for Appropriate Technology (NCAT) Information Staff
P.O. Box 3838
Butte, Mont. 59701

* Courtesy of *Solar Age*, SolarVision, Inc., Harrisville, New Hampshire.

New Mexico Solar Energy Association (NMSEA)
P.O. Box 2004
Santa Fe, N.M. 87501

W.W. Norton and Co., Inc.
500 Fifth Ave.
New York, N.Y. 10036

Portola Institute and Dell Publishing Inc.
1 Dag Hammerskjold Plaza
New York, N.Y. 10017

Princeton University Press
41 William St.
Princeton, N.J. 08540

Rodale Press
Organic Park
Emmaus, Penn. 18049

Schocken Books, Inc.
200 Madison Ave.
New York, N.Y. 10016

Sierra Club Books
530 Bush Street
San Francisco, Calif. 94108

Solar Usage Now
450 E. Tiffin St.
Bascom, Ohio 44809

SolarVision, Inc.
Church Hill
Harrisville, N.H. 03450

Total Environmental Action, Inc. (TEA)
Church Hill
Harrisville, N.H. 03450

Unipub
345 Park Avenue, South
New York, N.Y. 10010

University of Oregon, The Center for Environmental Research
c/o **The School of Architecture and Allied Arts**
Eugene, Ore. 97403

Van Nostrand Reinhold Co.
135 West 50th St.
New York, N.Y. 10020

Vermont Crossroads Press
Box 333
Waitsfield, Vt. 05673

Volunteers in Technical Assistance (VITA) Publications
3706 Rhode Island Avenue
Mt. Rainier, Md. 20822

The Whole Earth Truck Store
c/o The Zen Center
300 Page Street
San Francisco, Calif. 94102

John Wiley & Sons, Inc.
605 Third Ave.
New York, N.Y. 10016

Windbooks
Box 14C
Rockville Centre
New York, N.Y. 11571

Yale University Press
382 Temple Street
New Haven, Conn. 06520

Zomeworks Corporation
P.O. Box 712
Albuquerque, N.M. 87103

Organizations and Agencies *

A. Regional Solar Energy Centers

Young but growing up fast, the regional solar energy centers are becoming muscular and muscle-bound (in federal paperwork) organizations with the promise of hastening the transition from fossil fuels to renewable energy in the American economy. National headquarters for solar energy information and development are at the Solar Energy Research Institute (SERI) in Golden, Colo. It has mushroomed in the last 18 months and can now offer a dizzying supply of services to solar professionals and solar organizations — and it has a fabulous solar library. Both SERI and the four regional centers circulate newsletters

---

* Courtesy of *Solar Age*, SolarVision, Inc., Harrisville, New Hampshire.

describing projects, services and publications. Get yourself on the mailing list at SERI and at your regional center. A visit to the libraries and information offices of each center can also be profitable. Much of the assistance available from SERI and the regional centers is also channeled through state solar energy offices and federal information dissemination agencies, many of which are mentioned in this directory.

Solar Energy Research Institute (SERI)
1617 Cole Blvd.
Golden, CO 80401
(303) 231-1415 (reference desk number)

Northeast Solar Energy Center
470 Atlantic Avenue
Boston, MA 02110
(617) 292-9250
(Serves: Connecticut, Maine, Massachusetts, New Hampshire, New Jersey, New York, Pennsylvania, Rhode Island, Vermont)

Mid-American Solar Energy Complex
8140 26th Avenue, South
Bloomington, MN 55420
(612) 853-0400
(Serves: Illinois, Indiana, Iowa, Kansas, Michigan, Minnesota, Missouri, Nebraska, Ohio, North Dakota, South Dakota, Wisconsin)

Southern Solar Energy Center
61 Perimeter Park
Atlanta, GA 30341
(404) 458-8765
(Serves: Alabama, Arkansas, Delaware, District of

Columbia, Florida, Georgia, Kentucky, Louisiana, Maryland, Mississippi, North Carolina, Oklahoma, Puerto Rico, South Carolina, Tennessee, Texas, Virgin Islands, Virginia, West Virginia)

Western SUN
715 Southwest Morrison
Portland, OR 97204
(503) 241-1222
(Serves: Alaska, Arizona, California, Colorado, Hawaii, Idaho, Montana, Nevada, New Mexico, Oregon, Utah, Washington, Wyoming)

B. National Organizations

Anyone seriously interested in solar energy will stay abreast of the activities of these organizations and obtain information about the publications they offer.

American Section of the International Solar Energy Society, Inc. (AS/ISES)
Advanced Institute for Research Technology
Highway 190 West
Killeen, TX 76541

Center for Energy Policy & Research
New York Institute of Technology
Old Westbury, N.Y. 11568
(516) 686-7578

Center for Renewable Resources
1001 Connecticut Ave., N.W.
Suite 510
Washington, D.C. 20036
(202) 466-6880

Citizens' Energy Project
1110 6th St., N.W.
Suite 300
Washington, D.C. 20009
(202) 387-8998

Institute for Ecological Policies
9208 Christopher Street
Fairfax, Va. 22031
(703) 691-1271

Institute for Local Self-Reliance
1717 18th Street, N.W.
Washington, D.C. 20009
(202) 232-4108

National Association of Home Builders (NAHB)
15th and M Streets, N.W.
Washington, D.C.
(202) 452-0271

**National Association of Solar
Contractors**
910 Seventeenth St., N.W.
Suite 928
Washington, D.C. 20006
(202) 785-3244

**National Center for
Appropriate Technology
(NCAT)**

P.O. Box 3838
Butte, Mont. 59701
(406) 494-4572

**Solar Energy Industries
Association (SEIA)**
1001 Connecticut Avenue,
N.W.
Suite 800

Washington, D.C. 20036
(202) 293-2981

Solar Lobby
1001 Connecticut Avenue,
N.W.
5th Floor
Washington D.C. 20036
(202) 466-6350

C. AS/ISES *

Chapters of the American Section of the International Solar Energy Society can be a rich source of information and services. You should get to know your local association and the professional help and published materials (including newsletters) and audiovisuals they offer:

AS/ISES CHAPTERS
**Alabama Solar Energy
Association**
Bernard Levine, Chairman
c/o University of Alabama
Center for Environmental
Energy Studies
UAH/JEEC RI Annex D
P.O. Box 1247
Huntsville, Alabama 35807
(205) 895-6257

**Arizona Solar Energy
Association**
Don Osborn, Chairman
(602) 255-3682
Buck Orndorff, Secy.-Treasurer
c/o FCAT
P.O. Box 1443
Flagstaff, Arizona 86002
(602) 779-3110 or 0505

**Colorado Solar Energy
Association**
Rachel Snyder, President

P.O. Box 5272
Denver, Colorado 80217
(303) 231-1192

**Eastern New York Solar Energy
Association**
William Rogers, President
(518) 270-6702
Dave Barton, Secy.
P.O. Box 5181
Albany, N.Y. 12205
(518) 270-6301

**Florida Solar Energy
Association**
Arthur Bowen, President
P.O. Box 248271
University Station
Miami, Florida 33124
(305) 284-3438

**Georgia Solar Energy
Association**
Tom McGowan, President

P.O. Box 32748
Atlanta, Georgia 30332
(404) 894-3448

**Hoosier Solar Energy
Association, Inc.**
Gordon Clark, Chairman
Gordon Clark Assoc., Inc.
6523 Carrolton Ave.
Indianapolis, Indiana 46204
(317) 259-7711

**Illinois Solar Energy
Association**
James Hartley, President
6 N 201 Denker Road
St. Charles, IL 60174
(312) 377-1509

Iowa Solar Energy Association
Roger P. Hadley, President
1433 Wildwood Drive, N.E.
Cedar Rapids, Iowa 52402
(319) 365-6103

---

* Courtesy of *Solar Age*, SolarVision, Inc., Harrisville, New Hampshire.

Kansas Solar Energy Association
Donald R. Stewart
1202 South Washington
Wichita, Kansas 67211
(316) 262-7427

Metropolitan N.Y. Solar Energy Association
William Bobenhausen,
President
P.O. Box 2147
Grand Central Station
New York, New York 10163
(914) 948-6490

Michigan Solar Energy Association
Karen Kanniainen, Coordinator
201 E. Liberty Street, Suite 15
Ann Arbor, Michigan 48104
(313) 663-7799

Mid-Atlantic Solar Energy Association
Linda Knapp, Director
2233 Grays Ferry Avenue
Philadelphia, Pennsylvania 19146
(215) 963-0880

Minnesota Solar Energy Association, Inc.
Larry Opseth, President
Myers Bennet Architecture
2829 University Ave., S.E.
Minneapolis, Minnesota 55414
(612) 379-7878

Mississippi Solar Energy Association
Dr. Pablo Okhuysen, Chairman
225 West Lampkin Road
Starkville, Mississippi 39759
(601) 323-7246

Nebraska Solar Energy Association
Dr. Bing Chen

University of Nebraska
Department of Electrical Technology
60th & Dodge Streets
Omaha, Nebraska 68182
(402) 554-2769

Nevada Association of Solar Energy Advocates
Shelly Gordon
P.O. Box 8179
University Station
Reno, Nevada 89507

New England Solar Energy Association
Susan Luster, Administrative Director
P.O. Box 541
Brattleboro, Vermont 05301
(802) 254-4221

New Mexico Solar Energy Association
Stephen Meilleur, Executive Director
P.O. Box 2004
Santa Fe, New Mexico 87501
(505) 983-1006/983-2861
983-2887

North Carolina Solar Energy Association
Ben T. Gravely, Chairman
7001 Buckhead Drive
Raleigh, N.C. 27609
(919) 566-3111 (o)
(919) 847-1188 (h)

Northern California Solar Energy Association
Harry Miller, President
120 Belgian Drive
Dansville, CA 94526
(415) 837-6381

Ohio Solar Energy Association
Dr. Bruce T. Austin, Chairman

Environmental Studies
Program Secretariat
Wright State University
Dayton, Ohio 45435
(513) 767-7324,
ext. 78/873-2169

Oklahoma Solar Energy Association
Dr. Bruce V. Ketcham
Solar Energy Laboratory
University of Tulsa
Tulsa, Oklahoma 74104
(918) 592-6000

Pacific Northwest Solar Energy Association
Dr. Douglas R. Boleyn,
Chairman
2332 E. Madison
Seattle, Washington 98112
(503) 226-8478

Tennessee Solar Energy Association
Mr. Joe Hultquist, Executive Director
P.O. Box 448
Jefferson City,
Tennessee 37760
(615) 397-2594

Texas Solar Energy Society
Russell E. Smith, Executive Director
1007 South Congress, Suite 359
Austin, Texas 78704
(512) 443-2528

Virginia Solar Energy Association
Dr. Robert Poignant
c/o Piedmont Technical Associates
300 Lansing Avenue
Lynchburg, Virginia 24503
(804) 846-0429

**Wisconsin Solar Energy
Association**
Michael Ducey, Chairman
1121 University Avenue
Madison, Wisconsin 53715
(608) 251-4447

Passive Solar Construction Assistance

A. Passive Products

1. *Solar Heat Collecting Greenhouses* *

Energy Shelters, Inc.
2162 Hauptman Rd.
Saugerties, NY 12477
Morton Schiff (914) 246-3135
Insulated shell with Beadwall
south glazing.

Solar Resources, Inc.
P.O. Box 1848
Taos, NM 87557
Stephen R. Kevin
(505) 758-9344
Twin skin, air-inflated
sunspace Solar Room™.

Solar Technology Corp.
2160 Clay St.
Denver, CO 80211
Richard Speed (303) 455-3309
Solera/solar garden greenhouse.

Vegetable Factory, Inc.
100 Court St.
Copiagne, NY 11726
John Stevens (516) 842-9300
Lean-to and freestanding solar
greenhouses.

2. *Insulating Shutters and Shades* †

Appropriate Technology Corp.
P.O. Box 975, 22 High St.
Brattleboro, VT 05301
May, David A.
Makes roll-up thermal shade
called Window Quilt.

Ark-Tik-Seal Systems, Inc.
P.O. Box 428
Butler, WI 53007

Restle, Joseph W.
Makes roll-up thermal shade.

**Center for Community
Technology**
1121 University Ave.
Madison, WI 53715
Korda, Randy, Executive
Director
Provides plans for multilayer
home-made shutters and
shades.

Conservation Concepts Ltd.
Box 376
Stratton, VT 05155
Moorehead, Claire F., President
Sells thermal curtain called
WARM-IN Sealed Drapery
Liner.

---

* Courtesy of *Solar Age*, SolarVision, Inc., Harrisville, New Hampshire.
† From *Thermal Shutters and Shades*. Courtesy of Dr. William Shurcliff, Brick
House Publishing Co., Andover, Massachusetts.

Green Mountain Homes, Inc.
Royalton, VT 05068
Kachadorian, Jim
Sells thermal shutters.

Insul Shutter, Inc.
Box 338
Silt, CO 81652
Sells thermal shutters for
windows and doors.

Insulating Shade Co., Inc.
P.O. Box 282
Branford, CT 06405
Carlson, Gustaf
Sells 3, 4, or 5-layer roll-up
shade assembly.

Rolscreen Co.
Pella, LA 50219
Stuart, Autyn
Makes Pella Rolscreen shutter.

Shutters, Inc.
110 E. 5 St.
Hastings, MN 55033
Swanstrom, P.W.
Has Patent 4,044,812 on hinged
folding shutter.

Therma-Roll Corp.
512 Orchard St.
Golden, CO 80401
Banbury, John Q., II, President

Makes outdoor roll-up type
shutter employing hollow slats.

Thermal Technology Corp.
P.O. Box 130
Snowmass, CO 81654
Shore, Ronald
Developed self-inflating
thermal curtain.

Zomeworks Corp.
P.O. Box 712
Albuquerque, NM 87103
Baer, Steven C.
Sells Beadwall, Skylid, and
other thermal shutter devices.

3. Glazing Manufacturers *

American Acrylic Corporation
173 Marine St.
Farmingdale, NY 11735
M. Ziegler (516) 249-1129
Lumasite Fiberglass Acrylic
Sheets

ASG Industries, Inc.
P.O. Box 929
Kingsport, TN 37662
M. L. Lilly (615) 245-0211
Low-iron crystal glass,
Sunadex™ Lo-Iron™
transparent, tempered sheet
glass

C-E Glass
Division of Combustion
Engineering, Inc.
825 Hylton Road
Pennsauken, NJ 08110
Dennis McKinny
(609) 662-0400
Heliolite™ low-iron pattern
glass.

CY/RO Industries
859 Berdan Ave.
Wayne, NJ 07470
(201) 839-4800
Acrylite SDP double-skinned
acrylic sheet and Polycarbonate
SDP double-skinned sheet

E.I. DuPont de Nemours &
Company, Inc.
Plastic Products and Resins
Department
Technical Service Laboratory
Room 420, Chestnut Run
Wilmington, DE 19898
Brian Mead (302) 774-5852
Teflon FEP film

Filon Division
Vistron Corporation
12333 Van Ness Avenue
Hawthorne, CA 90250
James E. Whitridge
(213) 757-5141
Filon™ Tedlar™ glazing, Type

548 and 558 acrylic-enriched
polyester panels

Helio Thermics, Inc.
110 Laurens Road
Greenville, SC 29607
William Haas (803) 235-8529
Helio Thermics Collector
Cover — two layers of cross-
corrugated fiberglass; top layer
is Tedlar coated

Kalwall Corporation
Solar Components Division
P.O. Box 237
Manchester, NH 03105
Scott Keller (603) 668-8186
Sunwall™ and Sun-Lite™

Park Energy Company
Star Route, Box 9
Jackson, WY 83001
Frank D. Werner (307) 733-4950
Glazing A2-3 through A2-9
Polycarbonate extruded into a
double glazing

* Courtesy of *Solar Age*, SolarVision, Inc., Harrisville, New Hampshire.

Rohm and Haas Company
Independence Mall West
Philadelphia, PA 19105
A. D. Jung (215) 592-3000
Tuffak-Twinwall℠
Polycarbonate Sheet

Sheffield Plastics, Inc.
P.O. Box 248
Salisbury Road
Sheffield, MA 01257
Thomas Kradel (413) 229-8711
Poly-glaz (polycarbonate)

Swedlow, Inc.
7350 Empire Dr.
Florence, KY 41022
Swedcast 300 acrylic

B. Passive Construction Plans

The following list of solar plans has been kept up-to-date by Mark Hopkins at the National Center for Appropriate Technology in Butte, Montana and by *Solar Age*, a magazine published by SolarVision, Inc., Harrisville, New Hampshire.

1. *Greenhouse/Sunspace*

Ecotope
2332 E. Madison Street
Seattle, Washington 98112

Total Environmental Action Foundation
Church Hill
Harrisville, NH 03450
Plans available with consultation.

Domestic Technology Institute
Box 2043
Evergreen, Colorado 80439
An Attached Solar Greenhouse

The Lightning Tree
P.O. Box 1837
Sante Fe, NM 87501
Beadwall System Window and Greenhouse Plans

Zomeworks Corporation
P.O. Box 712
Albuquerque, NM 87103
Solar Greenhouse Performance and Analysis

University of Oregon
Department of Agriculture
Eugene, OR 97403
Solar Frame Plans

Solar Survival
P.O. Box 119
Harrisville, NH 03450
Solar Greenhouse Construction Drawings

Farallones Institute
1916 5th Street
Berkeley, CA 97710
A Solar Greenhouse for Mesa College

Grand Junction Public Energy Information Office
250 North Station
Grand Junction, CO 81501
Vocational Region 10 Solar Greenhouse

Maine Audubon Society
118 Old Route One
Falmouth, ME 04105

Solar Applications & Research Ltd.
3683 West 4th Avenue
Vancouver, B.C. Canada V6R 1P2

Solar greenhouse stock plans with siting and operating guide $35 plus 1.50 postage and handling.

Sunspaces
2520 Sacramento Street
San Francisco, CA 94115
(415) 922-0382
Solar Greenhouses for Mobile Homes
A manual for mobile homeowners who live in sunny but cold regions.
$2.75 postpaid.

South Dakota Office of Energy Policy
State Solar Office
Capitol Lake Plaza
Pierre, SD 57501
A Solar Greenhouse for Your Mobile Home
Does not physically attach to the home or deface the structure, but does link with the home through an existing window. Materials for the design cost from $500–$550 at 1979 prices.

2. *Solar Collectors and Systems*

Akira Kawanabe, Architect
P.O. Box1014
1417 Main Street
Alamosa, CO 81101
(303) 589-4336
A packet of plans is available for
$1.00. Responsible for
approximately 15 installations,
and for many more through
workshops. Offers complete
design services.

**Ayer's Cliff Centre for Solar
Research**
P.O. Box 334
Ayers Cliff
Quebec
Canada J0B 1C0
(819) 838-4871
Plans ("Prototype Canada")
available for $8.00. An update
paper on current trends can
be obtained for 75¢ (mailing and
handling). Will act as liaison
between homeowners and
builders. Has designed and built
approximately ten systems and
designed and supervised eight
others.

Bio-Energy Systems, Inc.
Mountaindale Road
Spring Glen, NY 12483
(914) 434-6322
Offers a manual describing
how their product (Solaroll) can
be used.

**Total Environmental Action
Foundation**
Church Hill

Harrisville, NH 03450
Passive Air Panels (T.A.P.)
(Wall mounted. Plans
available with
consultation.)

The Solar Project
630 Rockland Street
Lancaster, PA 17602

Midland Energy Institute
1221 Baltimore Street
Kansas City, MO 64106
Fan-assisted air panels.
(Wall mounted.)

Domestic Technology Institute
Box 2043
Evergreen, CO 80439

**San Luis Valley Solar Energy
Association**
Box 1284
Alamosa, CO 81101

**Illinois Institute of Natural
Resources**
325 West Adams Street
Springfield, IL 62706
*Solar Space Heaters for Low-
Income Families* by Roger
Fenton and Patti Donahue.
Illinois Valley Economic
Development Corporation. A
manual designed to guide
individuals in the construction
and installation of a low-cost
window box solar collector.
Includes a bibliography of
information on home
weatherization and solar
energy use.

J. K. Ramsetter
1520 West Alaska Place
Denver, CO 80223
(303) 733-8552
Plans are available for $10.00
(an estimated 500 installations
have been built from them).
Design Services. Approxi-
mately 130 installations in
Colorado.

Rapp Solar Design, Inc.
29 Highgate Road
Marlboro, MA 01752
Plans for the "2-pass" site-built
collector are available this
summer. (Diy-Sol, Inc., at the
same address, manufactures
the absorber plate for this
collector.) Offers design
services. Approximately
twenty installations.

Solstice Publications
Box 2043
Evergreen, CO 80439
"Solar Forced Air Heating
System" plans for $7.50.

**Total Environmental Action,
Inc.**
Church Hill
Harrisville, NH 03450
(603) 827-3374
Plans with manual for the
MODEL-TEA collector
(designed by TEA, Inc. under
contract with the U.S.
Department of Energy) is
available. Complete design and
engineering services.
Approximately ten site-built
installations.

3. Other Plans

John Bloodgood, Architects
3001 Grand
Des Moines, Iowa 50312
Four passive solar house plans
$75/one set, $150/eight sets.
$195/Mylar Sepias. Plans are for
a single level ranch, townhouse,
zero-lot-line, and two-story
detached house.

Ecos Corporation
New Britain Avenue
P.O. Box 331
Farmington, CT 06032
James M. Salisbury
(203) 673-2110
Plans available for homes,
offices, and light commercial
structures.
Ecosolon™—a wide variety of
contemporary structures,
which include provisions for
optional active solar space
heating systems.
Ecoviron™—cluster concept of
sun-tempered special units.
Ecoterron™—earth-sheltered
structures.

Helio Thermics, Inc.
Donaldson Industrial Park
1070 Orion Street
Greenville, SC 29605
(803) 299-1300
Plans are available for $15.
They sell the glazing, and
manufacture and sell air
handler and controls. Ask about
their flat roof collector. Provide
design services. Approximately
80 installations completed.

R.G. Huber Construction
1009 Gregory
Normal, IL 61761
(309) 452-7771
Offers energy efficient and solar
oriented house plans.

Klaus Linnemayer
Natural Energy Architecture
P.O. Box 215
Holderness, NH 03245
(603) 968-3141
Offers eight designs for passive
homes. A new design $500 for a
set of 10 plans; seven other
designs $75/one set, $110/four
sets, $150/eight sets; $175/
Mylars.

**New York State Energy
Research and Development
Authority**
Rockefeller Plaza
Albany, NY 12223
NYSERDA award-winning
passive solar house designs.
Available for reproduction
costs.

**Total Environmental Action,
Inc.**
Church Hill
Harrisville, NH 03450
(603) 827-3374
Goosebrook Solar Home.
Plans are available for a 1,585
square foot, three bedroom
house complete with passive
solar heating features.

Depending on the type of active
solar system installed, fuel bills
in the house can be less than
one-fifth those of most home
heating bills (Active system
plans not included) $75 eight
construction sheets and bill of
materials. $10 extra sets if
ordered at the same time.

**Total Environmental Action
Foundation, Inc.**
Church Hill
Harrisville, NH 03450
(Plans available with
consultation) Additional Direct
Gain Glazing

Small Home Council
University of Illinois
Champaign-Urbana, IL 61801
Details and engineering
analysis of Illinois Lo-Cal
House.

**Dept. of Mechanical
Engineering**
University of Saskatchewan
Saskatoon, Saskatchewan
Canada S7N 0W0
An air-to-air heat exchanger for
residences.

Vermont State Energy Office
State Office Building
Montpelier, VT 05602
*Energy Efficient Retrofits for
Mobile Homes.* A four-page
guide offering the mobile
homeowner hints to take
advantage of the sun's heat.
Limited free distribution.

C. Other Passive Resources

David Bainbridge
The Passive Solar, Inc.
P.O. Box 7722
Davis, CA 95616
Passive Solar Catalog

Jeremy Coleman
Box 45
Marlboro, VT 05344
(802) 257-0735 or 7644
No plans available. Provides design and installation service. Approximately eight installations.

Contemporary Systems, Inc.
Walpole, NH 03608
(603) 756-4796
Manufactures and sells collector plates and air movers for site built systems. Design and engineering services. Ductwork and rock storage experts.

Dalen Products, Inc.
201 Sherlake Dr.
Knoxville, TN 37922
E. Neal Caldwell
(615) 690-0050
Self-operating solar vents.

Ekose'a
573 Mission Street
San Francisco, CA 94105
Lee Porter Butler
(415) 543-5010
Natural energy conserving architecture, design development, construction marketing, especially "solar envelope" houses.

Earthworks
Bucksport, ME 04416
(207) 469-7494
No plans are available, but as architects and builders they can help with all aspects of site-building. Approximately fifteen installations.

Hawkweed Group
4643 N. Clark Street
Chicago, IL 60640
(312) 784-5025
No plans available. An architecture and planning firm with approximately fifty site-built installations completed.

Skytherm Thermal Processing and Engineering
Harold Hay
2424 Wilshire Blvd.
Los Angeles, CA 90057
Inventor of Skytherm solar roofs and the only real expert on the subject.

Hot Stuff
P.O. Box 306
LaJara, CO 81140
(303) 274-4069
Sell solar components for site-built collectors with spec sheets on systems. Putting together a manual on all types of solar systems. Approximately 10 installations completed.

Kalwall Corp.
Solar Components Div.
P.O. Box 237
Manchester, NH 03105
Scott F. Keller (603) 668-8186
Fiberglass-reinforced polyester water storage tubes and other passive products.

One Design, Inc.
Mt. Falls Rd.
Winchester, VA 22601
Tim Maloney

(703) 877-2172 or 2196
Fiberglass-reinforced polyester waterwall modules.

Park Energy Company
Box SR9
Jackson, WY 83001
(307) 733-4950
Sells materials for their site-built system. Package is free to architects and their solar designers and $3.00 to others. Two types of collectors: 80-11 (no wood used in construction) and 80-15 (uses wood). Are primarily distributors and installers but can recommend design service people. Approximately fifteen installations.

Passive Solar Products Association
350 Endicott Bldg.
St. Paul, MN 55101

Solar Design Associates
Conant Road
Lincoln, MA 01773
(617) 828-7115
No plans, but offer complete design service and are experienced in installation. Have completed approximately five site-built installations.

Solstice Publications
Box 2043
Evergreen, CO 80439
Do-it-yourself blueprints, including an attached greenhouse, one- and-two-story free standing solar reliant greenhouse, large commercial passive solar greenhouse, passive solar grow hole,

moveable insulation system, solar forced-air heating system, solar food dehydrator and solar domestic hot water system. Also, passive solar house plans, including an adobe three-zone passive house and a low-cost urban passive solar house. Write for a free brochure listing plans and costs.

Woodenstone Construction Co.
c/o Woodenstone Acres Farm
Weare, NH 03281
No plans, but offers complete architectural and engineering services. Approximately five installations, including retrofits.

Wood, Wind & Sun
P.O. Box 146
Bolton, MA 01740
(617) 799-5533
Offers design services.
Experienced in site-building.

Solar Legislation

Federal Tax Incentives

The United States Congress has provided for substantial federal tax credits to encourage the individual to invest in energy-conserving measures and alternative energy sources. These tax breaks can result in substantial savings for solar investments. For more information order publication 903 from your local Internal Revenue Service.

A fact sheet published by the *Solar Lobby*, Washington, D.C., follows for a more complete understanding of what energy tax credits can do for you.

Fact Sheet On The Residential Federal Energy Tax Credits *

Congress has approved tax incentives to encourage energy conservation and development of renewable energy sources. This fact sheet tells how these tax credits work, what items are eligible, and how you as a taxpayer can go about claiming a credit if you are entitled to one.

What Is A Tax Credit?

A tax *credit*, you will find when preparing your income tax return, is much better than a tax *deduction*. Deductions are subtracted from your gross income, while a credit is subtracted directly from your tax bill. This means, for

---

* Courtesy *Solar Lobby*, 1001 Connecticut Avenue, N.W., Washington, DC 20036.

example, that if you are in a 20% tax bracket, a $1,000 *deduction*, would lower your tax payment by $1,000 × 20% or $200. A $1,000 *credit*, by contrast, would lower your taxes by $1,000.

What Are The Residential Energy Tax Credits?

There are two distinct residential energy tax credits, each with its own conditions and limits. The credit for *energy conservation* costs is 15 percent of any amount up to $2,000 (making the maximum credit $300). The credit for *renewable energy* costs (solar, geothermal, or wind-powered equipment) is 40 percent of any amount up to $10,000 (making the maximum credit $4,000).

The credits will be available for costs incurred through December 31, 1985.

The credits are not refundable. In other words, if the tax credit you are supposed to get totals more than the amount of taxes you owe, you do *not* get a refund for the difference. However, you *can* apply the excess credit against taxes in future years through December 31, 1987.

Who May Claim These Credits?

If you own *or rent* your home, you are eligible for the credits if (1) you pay for the qualifying equipment or materials and (2) the equipment or materials are used in your principal residence. The credits do not apply to equipment or materials used in vacation or second homes.

Landlords are *not* eligible for the credits. They are, however, eligible for the business investment energy tax credits (not covered in this fact sheet).

If you are a stockholder in a cooperative housing corporation or an owner of a condominium unit, you can claim a credit for your share of the cost of qualifying items which are installed by the corporation or condominium management association for the benefit of common users. (If you have an item installed in your individual unit at your expense, you will be eligible in the same manner as a homeowner or renter.)

In cases where qualifying renewable energy and conservation equipment is *jointly owned by two or more homeowners*, IRS considers each owner's investment separately, so each individual owner is eligible to take up to the maximum $4,000 credit for *their share* of the expense for their own dwelling.

If you *do business from your home*, you get the full credit only if less than 20 percent of the use of the qualifying item is for business purposes. If more than 20 percent is for business, you get a credit only for the percentage which is not used for business.

As for *capital gains*, if you own a house or condominium unit, its value will increase by the amount you spend on the item *less* the amount of the tax credit you receive. This means that when you sell the house and figure capital gains, you will not be able to use the amount of the credit to make the gain smaller.

With respect to *other energy subsidies and loans*, after January 1, 1980, you must choose between (1) the tax credits and (2) any other *subsidized energy loans* or *non-taxable grants*. However, *taxable grants* and *state and local income tax credits* can be taken in addition to the federal credits.

What Items Qualify For The Energy Conservation Credit?

The energy conservation items you install must be: (1) *new*, and (2) *purchased after April 19, 1977*. They must also be designed to last at least three years. (However, IRS has not yet developed standards for judging durability, and until they do, you may use your own judgment on this question.)

An item qualifies when its *installation at your principal residence is completed*. If you change residences, your new home can be considered your principal residence 30 days before you start living in it.

The full $2,000 worth of equipment covered by the credit does not have to be installed in a single tax year. That is, you can install, say, $500 worth in one year and $1,500 in the next, before exhausting your eligibility. Also, each principal residence you live in during the period of the credit's existence has its own separate $2,000 limit.

Expenses for the following items or measures are eligible for this credit:

a. *insulation* to reduce heat loss or gain in a home or water heater;

b. a *new furnace* designed to cut fuel consumption;

c. *flue restrictors* designed to increase the efficiency of the heating system;

d. an electric or mechanical *furnace ignition system* that replaces a gas pilot light;

e. a *storm or thermal door or window*;

f. an automatically timed *thermostat*;

g. *caulking or weatherstripping* an exterior door or window;

h. a *meter* that displays the cost of energy used;

i. other energy-saving items specified by the Secretary of Energy.

Expenses for the following items are *not* eligible for the credit: heat pumps; fluorescent lights; wood or peat stoves; replacement furnaces or boilers; hydrogen-fueled equipment; insulation that is primarily structural or decorative–carpets, drapes, wood paneling, exterior siding, etc.

What Items Qualify For The Renewable Energy Credit?

The renewable energy items you install must be: (1) *new*, and (2) *purchased after December 31, 1979.* They also must be designed to last at least five years. (However, IRS has not yet developed standards for judging durability, and until they do, you may use your own judgment on this question.)

An item qualifies when its *installation at your principal residence is completed.* If you change residences, your new home can be considered your principal residence 30 days before you start living in it.

The full $10,000 worth of equipment covered by the credit does not have to be installed in a single tax year. Also, each principal residence you live in during the period of the credit's existence has its own separate $10,000 limit.

Expenses for the following items are eligible for the credit:

a. solar energy equipment for heating or cooling the home, for providing hot water for use within the home, or for generating electricity (e.g., photovoltaics) for the home;

b. wind energy equipment for generating electricity or other forms of energy for home use;

c. geothermal energy equipment.

Not all types of solar energy equipment qualify. Both "passive" systems (which use the building's structure to collect and store the sun's heat) and "active" systems (which use mechanical methods to move heat from the collectors to a storage system) *are* eligible for the credit. However, those parts of a "passive" system that are structural *do not qualify*. Examples are as follows:

Items in a "passive" system which *would not* qualify: greenhouses; skylights; thermal storage walls which support a roof; south-facing windows.

Items in a "passive" system which *would* qualify: free-standing thermal storage walls; roof ponds; fans; movable insulation; thermosyphon systems. Also, solar panels installed as roofing qualify even though they are a structural component.

How Do I Claim the Residential Energy Credits?

You can claim the credits on your Form 1040. On the 1980 Form 1040, the credit was on line 45; this may or may not change in 1981.

Figure the amount of credit on Form 5965 (Energy Credits) and attach it to your return.

You *cannot* claim an energy credit on Form 1040A.

Any credits that have a lower number, or letter, on Form 1040 should be taken first, except for the three refundable tax credits, which are:

a. 31—taxes withheld from income;

b. 39—credit for excise tax on non-business uses of gasoline and lubricating oil;

c. 43—the earned credit.

What If I Need Further Information?

For further information about:

> how tax credits work
>
> which items are eligible for the residential energy tax credits
>
> how you can claim these credits

Contact your local IRS office (listed in the white pages of your phone book under United States Government).

B. Legislation—Summary by State

Many states have also enacted legislation that lowers the cost of an initial solar investment by providing tax incentives or exemptions from state revenues. When coupled with federal IRS incentives, solar systems become extremely attractive in most states.

This state-by-state legislative summary and the cost table were prepared by the *Solar Lobby*, Washington, D.C.

State By State Description Of State Income Tax Incentives For Solar Energy Equipment *

(States not listed do not yet have solar legislation)

Alaska: 10% fuel conservation credit up to $200 per individual or married couple for (1) insulation; (2) insulating windows; (3) labor for items 1 or 2; (4) solar, wind, tidal, or geothermal energy systems. Expires 12/31/82.

Arizona: 35% residential solar income tax credit through 1983 after which it declines 5% each year until 1989. Maximum credit $1000. Also, a 25% credit for residential insulation and ventilation devices. Max. credit $100. Expires 12/31/84.

Arkansas: Individuals may deduct the entire cost of solar heating and cooling equipment from their taxable income. The cost of other energy-conserving devices may also be deducted.

California: 55% residential solar income tax credit up to $3000 max. If a federal credit is claimed, the state credit is reduced by the amount of the federal credit.

Colorado: Individuals may deduct the cost of solar, wind, geothermal, or renewable biomass systems from taxable income. However, the state legislature has passed, and the Governor will apparently sign legislation replacing the deduction with a 30% tax credit up to $3000 for renewables and a 20% credit for conservation up to $400.

Delaware: $200 tax credit for solar energy devices designed to produce domestic hot water.

Hawaii: 10% credit on income tax for individuals who purchase solar energy devices that are placed in service by 12/31/81.

Idaho: Allows an income tax deduction for a solar heating/cooling or solar electrical system installed in the taxpayer's residence. The deduction equals 40% of the cost in the first year and 20% of the cost in each of the next 3 years; maximum deduction in any year is $5000.

---

* Courtesy *Solar Lobby*, 1001 Connecticut Avenue, N.W., Washington, DC 20036.

Indiana: 25% residential solar energy income tax credit up to $3000.

Kansas: 25% tax credit for residential solar energy systems up to $1000. Credit expires 7/1/83. However, there is a bill on the Governor's desk, which will be signed any day, that will raise the credit to 30%.

Massachusetts: Provides a state income tax credit of 35% or $1000, whichever is less, on the "net expenditures" of a solar system. The net expenditure equals the total expense minus any exemptions, credits, grants and/or rebates from any federal or state program.

Michigan: Income tax credit may be claimed for a residential solar, wind, or water energy device that is used for heating, cooling or electricity. For 1980 it is 25% for the first $2000 spent, and 15% for the next $8000.

Minnesota: 20% income tax credit up to $2000 for renewable energy source equipment. Expires 12/31/81.

Montana: Energy systems using non-fossil fuel installed in a taxpayer's residence before 12/31/81 are eligible for a tax credit of 10% for the first $1000 and 5% of the next $3000. If a federal tax credit is also claimed, the state credit is reduced to 5% of the first $1000 and 2½% of the next $3000.

New Mexico: 25% income tax credit for solar energy systems up to $1000.

North Carolina: 25% income tax credit for solar energy systems up to $1000.

North Dakota: Provides for an income tax credit for solar or wind energy devices. The credit is 5% per year for two years.

Ohio: A personal income tax credit of 10% of the cost of a solar, wind, or hydrothermal energy system up to $1000.

Oklahoma: 25% income tax credit for solar energy devices used to heat, cool, or furnish electrical or mechanical power up to $2000. Expires on 1/1/86.

Oregon: 25% income tax credit for the installation of solar, wind, or geothermal energy systems up to $1000. Expires 1/1/85.

Vermont: 25% income tax credit for wood-fired central heating and solar or wind systems up to $1000. Expires 7/1/83.

Wisconsin: Provides a direct subsidy program to refund a portion of a solar energy system's cost. If the building on which the system is installed was on the local tax rolls before 4/20/77, the rate of the refund will be 24% in 1980. Expenses must exceed $500, but not exceed $10,000.

Actual Cost of Solar Hot Water or Space Heating System Taking Most * Federal and State Tax Incentives into Account

	$3000 Solar Hot Water System	$10,000 Solar Space Heating System
Alaska	$1,600	$5,800
Arizona	800	5,000
California	1,350	4,500
Colorado†	900	3,000
Delaware	1,600	6,000‡
Hawaii	1,500	5,000
Indiana	1,050	3,000
Kansas	1,050	5,000
Massachusetts	1,170	5,000
Michigan	1,150	4,300
Minnesota	1,200	4,000
Montana	1,700	5,875
New Mexico	1,050	5,000
North Carolina	1,050	5,000
North Dakota	1,500	5,000
Ohio	1,500	5,000
Oklahoma	1,050	4,000
Oregon	1,050	5,000
Vermont	1,050	5,000
Wisconsin§	1,080	3,600
Other States	1,800	6,000

* Not including system real estate tax exemptions or sales tax exemptions. Also, Arkansas and Idaho allow deduction of cost of system from taxable income, but are not mentioned.

† Colorado now offers a tax deduction, but shortly the 30% credit included in the above calculation will be signed into law.

‡ Delaware only offers the credit for domestic hot water systems.

§ Wisconsin has a refund system instead of a tax credit, as explained on adjoined documents.

Bibliography

Learn More About Passive Solar—
A Bibliography

Here are some of the best publications covering low-energy home design. The more you learn, the happier you are likely to be when you use the sun.

Anderson, Bruce, with Michael Riordan. *The Solar Home Book.* Andover, Massachusetts: Brick House Publishing Company, 1976.

Anderson, Bruce. *Solar Energy: Fundamentals in Building Design.* New York: McGraw-Hill, 1977.

Aronin, Jeffrey Ellis. *Climate and Architecture.* New York: Reinhold Publishing Corporation, 1953.

ASHRAE Handbook of Fundamentals. New York: American Society of Heating, Refrigerating, and Air Conditioning Engineers, 1967, 1972, and 1977.

Baer, Steve. *Sunspots.* Albuquerque, NM: Zomeworks Corporation, 1975.

Bahadori, Mehdi N. "Passive Cooling Systems in Iranian Architecture." *Scientific American*, Vol. 238, No. 2, February 1978.

Bainbridge, D.; J. Corbett; and J. Hofacre. *Village Homes' Solar House Designs.* Emmaus, Pennsylvania: Rodale Press, 1979.

Barnaby, C. S.; P. Caesar; and B. Wilcox. *Solar for Your Present Home.* Sacramento, California: California Energy Commission, 1977.

Bennett, Robert. *Sun Angles for Design.* Bala Cynwyd, Pennsylvania: Robert Bennett, 1978.

Bliss, Raymond W., Jr. "Atmospheric Radiation Near the Surface of the Ground: A Summary for Engineers." *Solar Energy*, Vol. V, No. 3, July/September 1961.

Clegg, P., and D. Watkins. *The Complete Greenhouse Book.* Charlotte, Vermont: Garden Way Publishing, 1978.

Climatic Atlas of the United States, Washington, D.C.: U.S. Department of Commerce, 1968.

Crowther, Richard, and Solar Group/Architects. *Sun/Earth.* Denver, Colorado: A. D. Hirschfield Press, Inc., 1976.

Daniels, G. *Solar Homes and Sun Heating.* New York, New York: Harper & Row, Inc., 1976.

Danz, Ernst. *Architecture and the Sun.* London: Thames & Hudson.

Davis, N. D., and L. Lindsey. *At Home in the Sun: An Open House Tour of Solar Homes in the United States.* Charlotte, Vermont: Garden Way Publishing, 1979.

Eccli, Eugene. *Low Cost Energy Efficient Shelter for the Owner and Builder.* Emmaus, Pennsylvania: Rodale Press, 1975.

Fisher, Rick, and W. Yanda. *The Food and Heat Producing Solar Greenhouse: Design, Construction, Operation.* Santa Fe, New Mexico: John Muir Publications, 1976.

Franta, Gregory E., and K. R. Olson. *Solar Architecture.* Proceedings, Aspen Energy Forum 1977. Ann Arbor, MI: Ann Arbor Science Publishers, Inc., 1978.

Geiger, Rudolf. *The Climate Near the Ground.* Cambridge, Massachusetts: Harvard University Press, 1950.

Givoni, B. *Man, Climate and Architecture.* Barking, Essex, United Kingdom: Applied Science Publishers, 1969.

Mazria, Edward. *The Passive Solar Energy Book.* Emmaus, Pennsylvania: Rodale Press, 1979.

McCullagh, J. C. *The Solar Greenhouse Book.* Emmaus, Pennsylvania: Rodale Press, 1978.

McPhillips, M. *The Solar Age Resource Book.* New York, New York: Everest House, 1979.

Morris, Scott W. "Natural Convection Collectors," *Solar Age*, Vol. 3, No. 9, September 1978.

Olgyay, Victor. *Design with Climate.* Princeton, New Jersey: Princeton University Press, 1963.

Olgyay, Aladar, and V. Olgyay. *Solar Control and Shading Devices.* Princeton, New Jersey: Princeton University Press, 1967.

Oddo, S. *Solar Age Catalog.* Harrisville, New Hampshire: SolarVision, Inc., 1977.

Shurcliff, William A. *Thermal Shutters and Shades.* Andover, Massachusetts: Brick House Publishing Company, 1980.

Shurcliff, William A. *Solar Heated Buildings of North America—120 Outstanding Examples.* Andover, Massachusetts: Brick House Publishing Company, 1978.

Spetgang, I., and M. Wells. *Your Home's Solar Potential.* Barrington, New Jersey: Edmund Scientific Company, 1976.

Wells, M., and I. Spetgang. *How to Buy Solar Heating . . . Without Getting Burnt.* Emmaus, Pennsylvania, 1978.

Wright, David. *Natural Solar Architecture: A Passive Primer.* New York, New York: Van Nostrand Reinhold Company, 1978.

Index

active systems, v
air conditioning. *See* cooling, insulation, ventilation.
air flow, for natural ventilation, 93
air stratification, 7
American Section of the International Solar Energy Society, Inc., (AS/ISES), listing of chapters, 174
attached sunspaces. *See* solar rooms.
audiovisuals, listing of sources, 171

backdraft dampers, 57, **58**
Baer, Steve C., 56, 61
    address of, 176
Balcomb residence, 80, **124**
Beadwall, 50
Bier, Jim, 63
Bliss, Raymond W., 95
Brookhaven House, 127, **122, 135–137**
Butler, Lee Porter, 82, **125**
    address of, 182

caulking and weatherstripping, 20, 21
climatic design. *See* regional architecture.
collectors, compared with windows, 42
collectors, heat output, 59
comfort, 23

comfort chart, 23, **23**
conduction, 6, 19
conservation. *See* energy conservation.
construction plans, listing for passive systems, 179–81
convection, 6, 11. *See also* ventilation.
convective loops. *See* solar chimneys.
cooling, 25, 26, 87–98
  conservation, 19
  fans, 20
Cool Pool, 96
cost, 114–18
Cryo-Acrylic, SDP™, 163–64

daylighting, 38
degree day, 7
degree days, **tables of**, **154–59**
design temperatures, **tables of**, **154–59**
direct gain, 11
Drumwall, 61, **61**

earth cooling. *See* earth pipes, ground cooling.
earth pipes, 98
Ekose'a, address of, 182
energy comparison, obtainable from oil, electricity, and gas, 112
energy conservation, 18–25. *See also* heat transfer principles.

envelope house. *See Butler, Lee Porter.*
evaporation, 87, 94–95
evaporative cooling, 7

Filon™, 161, 163
Flexiguard™, 162, 163

glazing, 7, 41, 160–64. *See also* windows.
  effect of several layers, 22, 43, 49, 75
  listing of manufacturers, 178–79
  with solar rooms, 74–75, **74**
  with solar walls, 67
  tables of types, **163**
greenhouses. *See* solar rooms
ground cooling, 96–98

Hay, Harold, 14, 71, 95
  address of, 182
heat capacity, 45
  table of, **45**
heat of fusion, 9
heat storage. *See* thermal mass
heat transfer principles, 6–9, **6–9**, 27–28. *See also* energy conservation.
house design, 34. *See also* length/width ratios, lot orientation.
  effect of passive on appearance, 5. *See also* regional architecture.

house design, *(continued)*
    for summer climates, 106–108
    for winter climates, 103–106

infiltration, 19
insolation. *See* solar gains
insulating shutters and shades, listing of suppliers, 177–78
insulation, 7, 20, 21–22. *See also* movable insulation, R-values.

Kalwall Corporation, address of, 182

landscaping, 36
    for enhancing natural ventilation, 91
length/width ratios, 37, **37**
Living Systems, 96
lot orientation, 34–35
mean radiant temperature, 6, 23
Mid-American Solar Energy Complex, address of, 174
movable insulation, 8, 11, 23, 28, 49–51, **50**. *See also* thermal window shades.
    energy savings from, 49
    with solar walls, 62
    table of U-values, 153

National Solar Heating and Cooling Information Center (NSHCIC), address of, 165
natural convection. *See* convection.
natural daylighting. *See* daylighting.
natural ventilation, 90–93
New Alchemy Institute, 78
night insulation. *See* movable insulation
Northeast Solar Energy Center, address of, 174

Olgyay, Victor and Aladar, 90
One Design, Inc., 64, **64**
orientation, 141. *See also* sun path diagrams.
overhangs, 88

passive solar energy, definition, v
plants, growth in solar rooms, 78
products, passive, listing of, 177–83
publishers, solar, listing of, 171

R-values, 9, 21–22. *See also* insulation.
radiation, 6, 87. *See also* sky radiation, solar gains.
    solar, **maps of sunshine percentages, 148–52**
    solar, tables of, 144–47
radiational cooling. *See* sky radiation.
regional architecture, 101–109
retrofitting, 20, 25–27. *See also* energy conservation.
roof ponds, 94. *See also* solar roofs.

shading, 8, 19, 25, 26, 87. *See also* solar control, overhangs, ventilation.
shading, of solar surfaces, 33
site, 34. *See also* landscaping, lot orientation.
site suitability, 26
sky radiation, 87, 95, 96
Skylid, 50
Skytherm, 14, 95
    address of, 182
social benefits, 4
Solar Age, address of, 171
solar attics, 26, 79, **79**
    retrofitting, 26
solar chimneys, 11–12, **12**, 28, **28**, 52–59
    air flow, 56–59

construction details, **54**
    for natural ventilation, 92, **92**
    retrofitting, 26
solar collectors. *See* collectors.
solar control, 87–90
solar conversions. *See* retrofitting.
solar cooling. *See* cooling.
Solar Energy Research Institute (SERI), address of, 174
solar gains, 31–32, 41, 43
Solar Lobby, address of, 175
solar position, 31–33, **31**
solar principles, 31–38
solar radiation, 32. *See also* solar gains.
solar retrofits, 26. *See also* retrofitting.
solar roofs, 14–15, **15**, 26, 70–72, **71**. *See also* solar attics.
solar rooms, 16, **16**, 29, **29**, 73–83
    all purpose design, 82
    listing of greenhouse suppliers, 177
    orientation of, 73
    retrofitting, 26
solar walls, 13, **13**, 29, **29**, 60–69
    construction of, 66–67, **66**
    for existing homes, 69
    summer shading of, 69
solar windows, 10, **10**, 28, **28**, 41–51. *See also* solar gains.
    compared with collectors, 42
Solatex™, 163, 164
Southern Solar Energy Center, address of, 174
specific heat, 45
    table of, **45**
state energy offices, listing of, 166–70
stratification. *See* air stratification.
SUEDE Project, **121**, **124**, 127, **129–134**, **138–140**
sun chart, 31
sun path diagrams, 141–43, **142–43**

Sunadex™, 163, 164
Sun-lite™, 161, 163

tax credits, 19
    for solar installations, 183–91
Tedlar™, 161–63
Teflon™, 161–63
thermal mass, 8, 44–48. *See also* heat capacity, specific heat.
    concrete floors, 48
    rules of thumb, 45
    use with solar rooms, 75
    use with solar windows, 46–48
thermal radiation. *See* radiation.
thermal storage roofs. *See* solar roofs.

thermal storage walls. *See* solar walls.
thermal window shades, 23
Total Environmental Action, Inc., address of, 180
Total Environmental Action Foundation, Inc., address of, 179
Trombe, Dr. Felix, 60
Trombe Wall. *See* solar walls.
Tuffak-Twinwall™, 163, 164

U-values, 9
    table for windows with insulation, **153**

vapor barriers, 22
ventilation, 87, 90. *See also* natural ventilation.

of solar rooms, 80
using earth pipes, 98
Vertical Solar Louvers, 63, **63**

water heating, 52
water walls, 61–66. *See also* solar walls.
weatherstripping. *See* caulking and weatherstripping.
Western SUN, address of, 174
windows, 7, 31–32. *See also* glazing, solar gains, solar windows.
    storm, 20, 22
    U-values of, 153
Wisconsin Solar Energy Association, address of, 177

Zomeworks Corporation, address of, 178

LEVEL 1

SCIENCE

LET'S READ AND FIND OUT

THUMP GOES THE RABBIT
HOW ANIMALS COMMUNICATE

BY FRAN HODGKINS · ILLUSTRATED BY TAIA MORLEY

HARPER

An Imprint of HarperCollinsPublishers

Special thanks to Dr. Fred Wasserman, Associate Professor of Biology
at Boston University, for his valuable assistance.

The Let's-Read-and-Find-Out Science book series was originated by Dr. Franklyn M. Branley, Astronomer Emeritus and former Chairman of the American Museum of Natural History–Hayden Planetarium, and was formerly co-edited by him and Dr. Roma Gans, Professor Emeritus of Childhood Education, Teachers College, Columbia University. Text and illustrations for each of the books in the series are checked for accuracy by an expert in the relevant field. For more information about Let's-Read-and-Find-Out Science books, write to HarperCollins Children's Books, 195 Broadway, New York, NY 10007, or visit our website at www.letsreadandfindout.com.

Let's-Read-and-Find-Out Science® is a trademark of HarperCollins Publishers.

Library of Congress Control Number: 2019935784
ISBN 978-0-06-249101-5 (trade bdg.) – ISBN 978-0-06-249097-1 (pbk.)
The artist used watercolor and traditional media with Adobe Photoshop to create the digital illustrations for this book.
Typography by Honee Jang
19 20 21 22 23 SPC 10 9 8 7 6 5 4 3 2 1

First Edition

To Tom—F.H.
For Maria Antonia, "you just gotta have fun."—T.M.

Birds sing. Cows moo. Dogs bark. Horses neigh. Animals use their voices a lot when they need to **communicate**.

But animals don't communicate only with their voices. Animals' ears, tails, feet, and bodies help them communicate, too. They can express everything from fear and anger to curiosity. Animals have a lot of ways to get their messages across.

The cat goes **RUB.**

Cats have special body parts called **glands** on their heads and sides. When a cat rubs up against something, the glands leave a scent behind. When other cats smell the scent, they get the cat's message: "This is where I am, and you're in my **territory**."

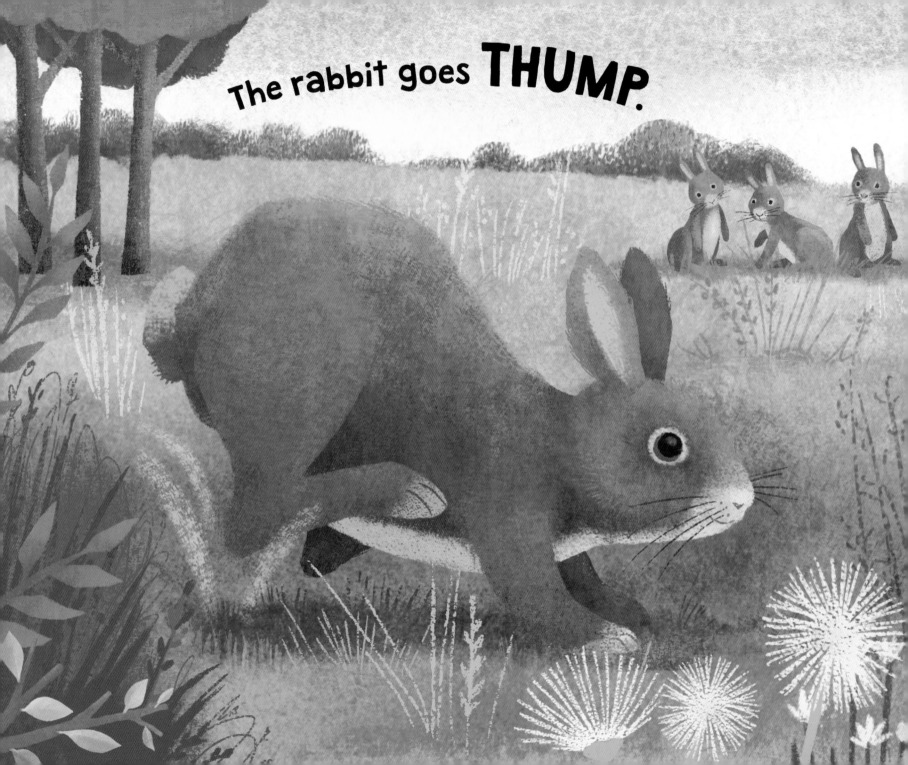

The rabbit goes **THUMP.**

Rabbits have long, strong back legs that are great for running fast and jumping high. They're also great for sending messages. The rabbit uses its strong legs to stomp its big back feet onto the ground. The stomp makes the ground vibrate. Other rabbits feel the **vibration**. The thump warns them, "Look out!"

The deer goes **FLASH**.

Deer are always alert for danger. If a deer is frightened, it will raise its tail and reveal a white patch on its hindquarters. Easy to see, the white patch tells other deer, "Run now!" It may also tell a **predator**, "Ha-ha! I see you and you can't catch me."

The firefly goes **FLICKER**.

16

On summer nights, ghostly lights flicker, some in the grass and some in the air. These flickering lights are fireflies. The males fly above the females, which are in the grass. They flash their lights to say, "Let's meet!"

The skunk goes STOMP.

A skunk's bold stripes act as a warning sign to other animals. But if those aren't enough, a skunk will stomp its front feet. The stomping says, "Better listen: This is your last chance before I spray!"

The fox goes **ROLL.**

When one fox meets another, sometimes they fight. But fighting is dangerous. So to avoid getting hurt, one fox may roll onto its back. When a fox shows its belly to another fox, it says, "I get it: You're in charge."

The rattlesnake goes **RATTLE**.

When it feels threatened, the rattlesnake
shakes the end of its tail. On the end are special
rings made of **keratin** (the same stuff that makes
up our fingernails). The rings make a dry clatter
that warns, "Go away, or I'll strike."

The whale goes SPLASH.

A 40-ton whale leaps out of the water and smashes back down, sending up a geyser of spray. Scientists call this behavior **breaching**. Some scientists think that a breaching whale is sometimes telling other whales, "Here I am!" But other scientists think whales breach for different reasons. Can you think of more reasons whales might breach?

The elephant goes **FLAP.**

26

Flap, flap, flap—elephant ears are useful for cooling off on a hot day, but they're more than just built-in air conditioners. They help an elephant express itself. Fast flaps mean excitement. Flapping ears and a raised head mean, "Hello!"

The owl goes **POOF.**

If you are big, other animals will leave you alone. An owl fluffs out its feathers and spreads its wings when it feels threatened. Sometimes, looking big is just as good as being big.

29

The bee goes **WAGGLE**.

A bee waggles for a very important reason: to tell other bees it has found food. Its waggling dance tells other bees which way and how far off the food is. This way, many bees can help bring back food that would be too much for one bee to carry.

The dog goes **wag**.

A dog's tail can tell a lot. When the tail is tucked between the dog's legs, the dog is frightened. When the tail is low and wagging a little, the dog is worried. But when a dog holds its tail high and wags it fast, that's an excited dog!

Furred, finned, or feathered, animals have many ways to communicate. How do animals communicate with you?

33

BE A CITIZEN SCIENTIST: MAKE OBSERVATIONS

You can make observations just like a scientist does. All you need is patience. Sit quietly near an animal. It could be outside, where you can watch birds at a feeder, or indoors, where you can watch your pet dog or cat. (Your pet might try to get you to play at first, but if you keep sitting still, your pet will go back to what it was doing.) Watch.

What does the animal do? When does it do it?
You can write down what you see.

Your observations will help you understand animals.

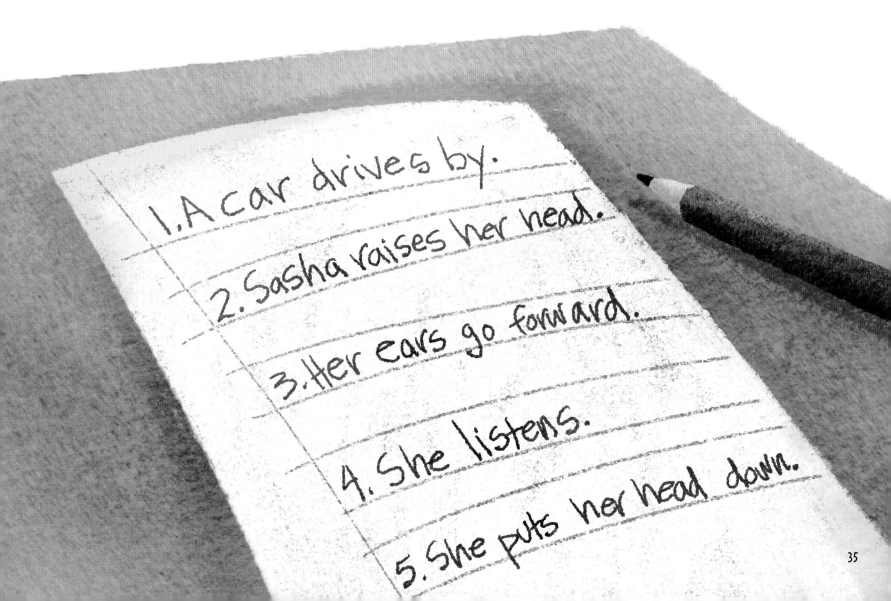

1. A car drives by.
2. Sasha raises her head.
3. Her ears go forward.
4. She listens.
5. She puts her head down.

GLOSSARY

Breaching: jumping out of the water and splashing back down

Communicate: to use words or movements to express feelings or give information

Glands: body parts that make a scent the body uses or gives off

Keratin: a flexible material that makes up hair and fingernails

Predator: an animal that hunts other animals

Territory: the area where an animal lives

Vibration: a fast back-and-forth movement

COMMUNICATION

LOOK OUT! =  =

LEAVE ME ALONE! =  =

HELLO! =  =

Be sure to look for all of these books in the Let's-Read-and-Find-Out Science series:

The Human Body:
How Many Teeth?
I'm Growing!
My Feet
My Five Senses
My Hands
Sleep Is for Everyone
What's for Lunch?

Plants and Animals:
Animals in Winter
The Arctic Fox's Journey
Baby Whales Drink Milk
Big Tracks, Little Tracks
Bugs Are Insects
Dinosaurs Big and Small
Ducks Don't Get Wet
Fireflies in the Night
From Caterpillar to Butterfly
From Seed to Pumpkin
From Tadpole to Frog
How Animal Babies Stay Safe
How a Seed Grows
A Nest Full of Eggs
Starfish
Super Marsupials
A Tree Is a Plant
What Lives in a Shell?
What's Alive?
What's It Like to Be a Fish?
Where Are the Night Animals?
Where Do Chicks Come From?

The World Around Us:
Air Is All Around You
The Big Dipper
Clouds
Is There Life in Outer Space?
Pop!
Snow Is Falling
Sounds All Around
The Sun and the Moon
What Makes a Shadow?

The Human Body:
A Drop of Blood
Germs Make Me Sick!
Hear Your Heart
The Skeleton Inside You
What Happens to a Hamburger?
Why I Sneeze, Shiver, Hiccup, and Yawn
Your Skin and Mine

Plants and Animals:
Almost Gone
Ant Cities
Be a Friend to Trees
Chirping Crickets
Corn Is Maize
Dolphin Talk
Honey in a Hive
How Do Apples Grow?
How Do Birds Find Their Way?
Life in a Coral Reef
Look Out for Turtles!
Milk from Cow to Carton
An Octopus Is Amazing
Penguin Chick
Sharks Have Six Senses
Snakes Are Hunters
Spinning Spiders
What Color Is Camouflage?
Who Eats What?
Who Lives in an Alligator Hole?
Why Do Leaves Change Color?
Why Frogs Are Wet
Wiggling Worms at Work
Zipping, Zapping, Zooming Bats

Dinosaurs:
Did Dinosaurs Have Feathers?
Digging Up Dinosaurs
Dinosaur Bones
Dinosaur Tracks
Dinosaurs Are Different
Fossils Tell of Long Ago
My Visit to the Dinosaurs
Pinocchio Rex and Other Tyrannosaurs
What Happened to the Dinosaurs?
Where Did Dinosaurs Come From?

Space:
Floating in Space
The International Space Station
Mission to Mars
The Moon Seems to Change
The Planets in Our Solar System
The Sky Is Full of Stars
The Sun
What Makes Day and Night
What the Moon Is Like

Weather and the Seasons:
Down Comes the Rain
Droughts
Feel the Wind
Flash, Crash, Rumble, and Roll
Hurricane Watch
Sunshine Makes the Seasons
Tornado Alert
What Makes a Blizzard?
What Will the Weather Be?

Our Earth:
Archaeologists Dig for Clues
Earthquakes
Flood Warning
Follow the Water from Brook to Ocean
How Deep Is the Ocean?
How Mountains Are Made
In the Rainforest
Let's Go Rock Collecting
Oil Spill!
Volcanoes
What Happens to Our Trash?
What's So Bad About Gasoline?
Where Do Polar Bears Live?
Why Are the Ice Caps Melting?
You're Aboard Spaceship Earth

The World Around Us:
Day Light, Night Light
Energy Makes Things Happen
Forces Make Things Move
Gravity Is a Mystery
How a City Works
How People Learned to Fly
How to Talk to Your Computer
Light Is All Around Us
Phones Keep Us Connected
Running on Sunshine
Simple Machines
Switch On, Switch Off
What Is the World Made Of?
What Makes a Magnet?
Where Does the Garbage Go?